TOWARD DISTANT SUNS

Stackpole Books

With Artwork by Don Dixon

TOWARD DISTANT SUNS

T. A. HEPPENHEIMER

TOWARD DISTANT SUNS

Published by
STACKPOLE BOOKS
Cameron and Kelker Streets
P.O. Box 1831
Harrisburg, Pa. 17105

Published simultaneously in Don Mills, Ontario, Canada
by Thomas Nelson & Sons, Ltd.

Printed in the U.S.A.

Library of Congress Cataloging in Publication Data

Heppenheimer, T A   1947-
    Toward distant suns.

    Bibliography: p.
    Includes index.
    1. Space colonies. 2. Outer space—Exploration.
I. Title.
TL795.7.H47 1979   999   79-16857
ISBN 0-8117-1578-7

To Carol

We shall not cease from exploration
And the end of all our exploring
Will be to arrive where we started
And know the place for the first time.

T. S. Eliot, ''Little Gidding,'' The Four Quartets

CONTENTS

Contents

Foreword

It's a surprising fact that even great prophets of the Space Age—
Tsiolkowski, Goddard, and Arthur C. Clarke—underestimated how
quickly it would burst upon us and how profoundly it would change the path of our civilization. Even
now we see only the barest beginnings of the change in course that lies ahead.

I wonder why? Is it because our existence is so rooted in the two-dimensional experience that we
cannot wrench our minds to contemplate the third realistically? Whatever the reason, Dr. Heppenheimer's *Toward Distant Suns* helps open our eyes to the three-dimensional universe, and helps us
appreciate that it is there not just to look at, but to feel and explore and move into.

Slowly and cautiously, but still much faster than we dreamed in the pre-Sputnik era, we are
learning how to move, work, and live in space. The U.S.S.R., committed to a substantial continuing
space program, is extending every few months our knowledge of human performance and survival in
zero gravity. The U.S., after the long twilight period that followed the Apollo era, soon will regain
manned orbital capability in a big way with the operation of the space shuttle.

Exploring the Solar System with space probes, sending men to the Moon and back, and even
building small enclosed cities on Mars were ideas familiar to us, almost banal, thanks to the
imagination of science-fiction authors. Those ideas involved spaceflight, but they didn't change our
fundamental "planetary chauvinism," our inborn feeling that most human beings, even centuries
from now, would be born, live, and die on planetary surfaces.

Within less than a decade, a rapidly growing number of people has come to realize that our
destiny is quite different, and far richer in options. There were "precursors" to these new ideas more
than a half-century ago.

Konstantin Tsiolkowski, a self-taught scientist born in nineteenth-century Russia, perceived that
the physical essentials of human existence—sunlight, air, gravity, land area, agriculture—could all
be obtained in far greater quantity than on Earth, and much more controllably, within man-made
rotating mini-worlds in space. He saw the asteroids as the ultimate "mine" for materials to build
them.

Later, J. D. Bernal in England wrote that humanity might build spherical honeycomb structures in orbit, housing many people. In the 1960s Freeman Dyson suggested that an ultimate civilization might free itself from its parent planet, move out into space, and expand until it would intercept all the visible light from its star, remaining detectable only in the infrared. Dandridge Cole, in *Beyond Tomorrow,* wrote of hollowing out asteroids to form multilayered cities and countries in space.

When published, all these ideas seemed safely distant in the future; as written, most were free of numerical calculation and were accepted as imaginative philosophical speculation.

My own work, begun in 1969, was quite different in approach and, as it turned out, far more threatening to accepted, establishment thought. In consequence, it took fully five years before I could get it published in a reviewed scientific journal. Fortunately, I resisted the temptation to lose patience in those years and turn the work into science fiction.

The new and upsetting conclusions were that space colonies could be built relatively soon, within the limits of known engineering practice, with ordinary materials; that they could be very large, as much as a hundred square miles in land area; and that they could be, if desired, extremely earthlike.

With asteroidal materials as the resource base, space colonies could be built with total land area as much as three thousand times that of the Earth. The same logic pursued a little farther brought the conclusion that nearly every star in our galaxy could be, ultimately, the center of a human civilization living in earthlike conditions.

After the years of frustration, it brought great satisfaction to me that the new ideas, once out in the open, spread like wildfire. It's been an even greater satisfaction that hundreds of people with great talent and diverse interests have joined the effort since 1974, and have contributed their enthusiasm, their good ideas, and their hard work.

Since 1975 the character and direction of the research have changed markedly. Two NASA studies carried out under my direction (1976 and 1977) emphasized using available space hardware (specifically the shuttle) to achieve as quickly as possible a substantial production rate in space of useful products such as satellite power stations. The participants in those studies were mainly aerospace engineers and scientists, and their conclusions, in signed technical papers, were subjected to the review of their scientific peers. The results of those studies are in volume 57 of the AIAA's "Progress" series and in NASA SP-428.

Dr. Thomas Heppenheimer is one of the most talented and articulate of those who have come on board since 1974. His personal specialty is orbital mechanics, the calculation of trajectories. He has made an important contribution with detailed studies on the paths of lunar material leaving the Moon from a catapult to a destination in space.

Fortunately, he also has a lively gift for popular writing. I hope that you also will choose to work toward this new and exciting future, rich in possibilities; read and enjoy.

GERARD K. O'NEILL

Gerard K. O'Neill, professor of physics at Princeton University, directs research under NASA sponsorship. His book The High Frontier: Human Colonies in Space *won the Phi Beta Kappa Science Book Award and is now*

published in eight languages.

Acknowledgments

It is a pleasure to take note of my friends and associates who have helped me with this book. Peter Goldreich, Paul Castenholz, Dom Sanchini, and John Nuckolls granted interviews or gave critical reviews of text. Eric Hannah, Robert Salkeld, Michael Hart, Eric Jones, and Keith Miller have furnished me with unpublished manuscripts or have described their work to me. In addition, I have had valuable discussions and comments from Peter Glaser, Brian O'Leary, Gerard K. O'Neill, John Billingham, Mike Papagiannis, Jill Tartar, Dave Black, Mark Stull, Tom Hagler, Neville Motts, Jerry Ross, Eugene Shoemaker, Jim Oberg, and Phil Chapman.

Particularly valuable have been my occasional-to-frequent get-togethers with Mark Hopkins; with Gayle Pergamit, Phil Salin, and the rest of the Stanford Committee for Space Development; and with Carol Motts, Terry Savage, and the other members of OASIS (Organization for the Advancement of Space Industrialization and Settlement). I also have been glad to discuss my ideas with Eric Drexler, Gayle Westrate, Nancy Williamson, Don Davis, Dave Ross, Don Dixon, Ellene Levenson, Bob Rubin, Ronnie Ross, Anita Gale, and Charles and Jeri Shuford.

A number of people have helped me in securing artwork and source materials. Among these have been James K. Harrison, Charles Darwin, Gordon Woodcock, Bill Rice, Joyce Lincoln, Sima Winkler, Sue Cometa, Roselle Killingbeck, Chuck Gould, Sandy Henry, Vera Buescher, Bill Gilbreath, Dick Preston, Tom Hagler, Gayle Westrate, Don Davis, Ron Miller, Robert Salkeld, Carolyn Henson, Charles Biggs, Ed Bock, Lou Fattorosi, Alan Wood, Mike Ross, Louis Parker, Jack Bell, and Larry King. Also, Joan Tregarthen took care of some of my photoduplicating.

Don Dixon deserves special mention for his artwork of high quality, for his unfailing support, and for his ability to come through with what I needed. His secretary, Suzie Osemore, has also helped. My editor, Neil McAleer, throughout this work has gone to great lengths to be sure it would meet his standards. Rose Kaplan, our consulting editor, has also given valuable critiques.

And, in all this I have had the support and encouragement of Carol Wilson, whose commitment to excellence has been an unfailing source of good cheer.

Preface

From the stars has come the matter of our world and of our bodies, and it is to the stars that we will someday return. These comings and goings are the theme of this book.

Conventional wisdom holds that stars suitable for the origin of life are common, and that planets on which life has arisen are far from rare. This viewpoint is quite in keeping with the past five centuries of Copernicanism in astronomy, which have steadily removed mankind farther and farther from the center of things.

However, in recent years there have been the beginnings of a contrary view. This view holds that far from being commonplace in the Galaxy, life such as ours has resulted from a most improbable concatenation of events. If we then are not at the center of things, at least we may have the pleasure of contemplating that only by rare accident have we been fortunate enough to evolve; that one could search hundreds of thousands of stars without finding our like. Yet life need not be tied to the planets where it arises. Intelligent cultures, by inventing the arts of space colonization and star flight, can make their presence felt on a galactic scale.

We thus are led to think about space colonization. The core of this book is a series of chapters which set forth an agenda and program, a sequence of space projects leading to the building of the largest imaginable space colonies. This sequence is a "sixfold way," a group of six major themes or efforts in space. No one of them exists merely to pave the way for greater things; all can be justified and pursued in and of themselves. Yet the sum total of these efforts is to give us true space colonies.

We begin with the space shuttle, which will gain us a routine access to space and allow initial work in space construction, demonstrating anew the promise beyond our Earth.

Next, there is the orbiting construction platform. Supported by the shuttle, it will allow construction of large communications spacecraft. Its most exciting uses, however, lie in experiments that will show the way to the power satellite to gather solar energy. It is the powersat that will stand at the center of a truly large space effort.

The powersat has been conceptualized as a structure as large as Manhattan Island, yet weighing only as much as an aircraft carrier. Such powersats today are receiving increasingly serious attention in Washington, and the first may be built before century's end. The building of powersats constitutes the third major theme of the six.

The first powersats will be assembled from materials brought from Earth, but they will become more economic if they are built from lunar resources. The use of such lunar resources in powersats will compel us to develop far-ranging paths of space commerce, with over a thousand people living in space. This is theme four.

Next come the first true space colonies. They will be built to serve the powersat project, providing comfortable homes for the space workers and their families. Bit by bit, they will become less dependent on Earth, more self-sufficient.

Finally, when all has been reduced to well-understood practice, these initial colonies will expand, but only when it becomes possible to build new colonies as cities in space, whose land will be available to all.

And when we have gone down this sixfold path, when at last the space colonies have grown and prospered, we will return again to the question of our place in the Galaxy. From a space colony to a starship is but a short step; it is a matter of adding appropriate power supplies and thermonuclear engines. What we may hope to do in the next century or so, perhaps other stellar civilizations have had the opportunity to do for billions of years. On the one hand, locales where intelligent life has arisen are rare. On the other hand, once intelligence arises it may sweep across the Galaxy in an evolutionary moment, and the places of its colonies may be quite widespread. We do not understand at all well our place in the Galaxy, but we may speculate. Among the matters for our speculation is whether after all we may be the only advanced civilization in the Galaxy.

The answer to this question may well lie in the distant future, but space colonization, or at least the major efforts which will point in its direction, are a matter for the next few decades. These colonies are not here yet, but they will come; they have long been foreseen.

CHAPTER 1

The Planetary Hang-Up

In the beginning there were stars and the void, and together they formed the Galaxy. And the Galaxy was not without form, nor was it featureless, nor was there darkness upon its face. In this time of beginnings the Galaxy was already approaching an age of 10 billion years. It had a mature, well-developed form, with a central bright nucleus and a disk of stars, the brightest of which were bunched into long curving lanes.

Then as now, the Galaxy was unimaginably rich and varied, alive with the pulsing processes of astrophysics, continually stirred and changed by stellar upheavals. As in the sea with its ceaseless tides and waves, storms and winds, and profusion of living creatures that exist for their season, be it for mere days or for many years, so it is with that greater sea, the Milky Way. If we think of the heavens as unchanging, it is merely because our own season is not yet of galactic proportions. We measure our days by the turning of our planet, our lives by its motion around the sun, but the days of the Galaxy are measured by its own turning motion, which takes two hundred million years.

Nor was the Galaxy made up only of stars. A thin, rarefied dispersion of hydrogen pervaded throughout. To illustrate the awesome scale of the Galaxy, one need only compare the concentration of this gas on a mountaintop—where air grows thin and is hard to breathe but which still has ten million trillion molecules in each cubic centimeter—to that throughout most of the space that was then between the stars, which was at less than one atom per cubic centimeter. In places it gathered into patchy clouds, and there it was richer, denser, possibly reaching all of one hundred atoms per cubic centimeter. Yet even in such a rich cloud, a volume the size of Earth would have held only four hundred pounds of matter, less than can be packed into the trunk of a car. But so vast was the Galaxy that all its gas together may have amounted to the mass of thirty thousand trillion earths.

There were waves within the galactic sea of stars in those days like there are today, which are called spiral arms. They give form and beauty to photos of galaxies, but their significance is far greater than that. It was the spiral arms, the spiral structure of the Milky Way, that formed our sun. To say that spiral arms are like waves in the sea is indeed to describe a little of how they behave. Waves cross the ocean and wash up on shore, but the ocean itself does not flow to the beach. It remains in its basin, and the waves merely ripple its surface.

Similarly, spiral arms are produced by disturbances that pass across the starfields. These disturbances are known as density waves because they cause stars to cluster together more thickly, or with greater density, than in regions where these waves are absent. Density waves emanate from the core or nucleus of some galaxies, a bright and active region where millions of stars may be concentrated in small areas of space. These stellar concentrations tend not to be regular in form, but to gather in an elongated pattern known as a bar. Impelled by the rotation of the galaxy, the bar rotates, its gravity producing a spiral-shaped pattern of disturbances in the galactic disk. These disturbances influence the motion of stars, causing them to stay near the spiral pattern for fairly long periods of time before leaving. The resultant clusterings of many stars perpetuate these gravitational disturbances indefinitely, producing permanent spiral arms. These then are regions of gravitational attraction, which travel around a galaxy but do not actually transport the stars. Instead, the spiral arms form a pattern that attracts new stars as it advances, while other stars leave.

Spiral arms also attract masses of interstellar gas. Amid this gas are the rich, dense (by cosmic standards!) clouds, extending for tens or hundreds of light-years. These clouds contain the mass of ten thousand stars, yet they do not collapse under their own gravity. Their internal pressure supports them, just as a tire full of air under pressure will support the weight of a car. Ordinarily these clouds behave somewhat like clouds in the sky—they form, accumulate, break up, and are slowly pushed here and there upon the cosmic tides. Occasionally, though, such a cloud grows dense enough to approach the point where it would overcome this pressure, collapse of its own weight. When such gas enters a spiral arm, it may receive the extra compression which triggers this collapse. The cloud then compresses under its own gravity and breaks up into a myriad of smaller clouds. As these in turn collapse and compress, they form a shower of stars.

The largest of these are hot, bright, and blue, ranging up to thirty or fifty times the mass of the sun. They form so soon after the arm passes that they illuminate its position in space, marking its passage with their brilliance. But it is this very brilliance that is their doom. The hydrogen from which they form serves as fuel for their thermonuclear fires, and in the most luminous stars, it doesn't last long—a few million years at most. When most of the hydrogen is gone, a core of helium is left. The star then contracts, increases its internal temperature, and calls on the helium as a source of energy. The helium responds and undergoes nuclear fusion, releasing energy while forming still heavier elements.

By that time, though, the star is in deep trouble. It no longer possesses its once vast reserves of nuclear fuels; instead it faces an energy crisis. The heavy elements are not only poor sources of nuclear energy, they are difficult to ignite. This is particularly true of carbon. A massive star may fuse its helium to develop a core of carbon, which then heats and compresses as the star further contracts. But instead of igniting and fusing to release more energy, the carbon may detonate. The result is a stellar cataclysm, a supernova explosion. For a few weeks the exploding star will shine with the brilliance of the galaxy itself, hurling its matter outward, back into the primordial gas from which the

A spiral galaxy. Its spiral arms are bright with young stars, some of which will evolve and become supernovae. (Courtesy Hale Observatories)

When a supernova blazes up, its brilliance may exceed that of its entire galaxy. (Courtesy Don Dixon)

star originally formed. Yet in this stellar death there is the promise of new life. Amid the supernova gases are the heavy elements formed prior to and during the explosion, from which planets may form.

About 5 billion years ago there was a cloud of gas within the Milky Way, differing little from the many that have existed before or since, with one exception: It contained atoms that in time would create you and I. This was the protosolar cloud. We cannot say anything of its form, its extent, or its age, but it is believed that some 4.7 billion years ago, with formation of the Solar System still 150 million years in the future, a spiral arm passed through the cloud and triggered the collapse of a portion of it. This portion formed a massive star, which shone brilliantly for a brief season—10 million years—and then exploded. This explosion enriched other portions of the cloud with heavy elements formed within that star, but did not otherwise influence the cloud. In its massive and quiescent grandeur, the protosolar cloud continued to exist.

Then 100 million years after that explosion, perhaps 50 million years after its debris reached our vicinity, another spiral arm passed through our cloud. Again it formed a new generation of massive stars, one of which was only sixty light-years from what would become the Solar System. This star too evolved and exploded, producing still more heavy elements to enrich the protosolar cloud; but it did something more. Like the immense hydrogen bomb that in fact it was, the supernova detonation sent a shock wave coursing through the cloud. It was weaker than the spiral arm and would have had little effect, except that the explosion was so close—a mere 350 trillion miles or so. Just as the shock from a hydrogen bomb can cause a building to collapse, so this shock caused parts of the protosolar cloud to collapse.

The supernova sent tongues of gas penetrating deep into the protosolar cloud, seeding it with yet more of the elements of life, and in due time the gestation of Sun and Solar System was under way. The astrophysicist refers to these intrusive gaseous tongues as the result of a ''Rayleigh-Taylor instability'' arising from the interactions of masses of gas, but we may think of it as the cosmic kiss, the interstellar sexual act that conceived our world.

How can we know of such things? In the mid-sixties, scientists at Berkeley found traces of the gas xenon in some meteorites. There was an excess of the isotope xenon 129, which proved to have formed from radioactive decay of another isotope, iodine 129. Four heavier xenon isotopes were also in evidence: Xe-131, -132, -134, and -136. The proportions present showed they had formed from fission of a different radioactive element, plutonium 244. Since these radioactive isotopes decay at known rates (17 million years for the half-life of iodine 129 and 82 million years for plutonium 244), it was possible to determine that the plutonium and iodine were formed no more than 100 million years before the meteorites. These elements are found only in the cores of supernovae, so they thus represented telltale evidence that there had been just such a stellar explosion.

In 1969 a meteorite fell near the Mexican town of Pueblito de Allende. Known as the Allende meteorite, its fragments were rushed to laboratories at Caltech and the University of Chicago, which had sensitive equipment for the study of moon rocks. The key findings came out of Caltech, from a laboratory with the formal name of The Lunatic Asylum. Only a very good scientist could get away with such a name, and indeed its two most senior inmates, Gerald Wasserburg and Dimitri Papanastassiou, are so well regarded that their colleagues count it quite an honor if they can get an invitation to work there. One who got such an invitation was Typhoon Lee. It was Lee who detected excess quantities of the isotope magnesium 26 in small grains within Allende. This most likely was produced by decay of another radioactive isotope, aluminum 26, which also must have come from a

The dust and gas from a supernova eventually attenuates and merges with the interstellar medium. (Courtesy Hale Observatories)

supernova. In contrast to the longer-lived iodine and plutonium, aluminum 26 decays very rapidly; its half-life is 720,000 years. This means it could have formed no more than a few million years before the Allende meteorite. The most appealing explanation is that the same supernova that formed the aluminum also was a second supernova, which triggered the formation of the Solar System. The conclusion of science is that from delicate and subtle studies of tiny grains in meteorites we can see evidence for important events leading to the origin of worlds.

Yet perhaps there is more. Supernovae are rare; perhaps one star in a thousand will detonate in that fashion, turning the nearby skies to fire and incinerating its planets in titanic seas of flame. Even though supernovae can serve as agents of planets' birth, at the moments of explosion they are the fulfillment of the prophecy of the Bhagavad Gita: ''I am become Death, the destroyer of worlds.'' It is not the usual thing that star formation is triggered by a supernova. Is there in this then a uniqueness of our Solar System? Did this supernova, in triggering its formation, endow it even before its birth with additional quantities of heavy elements, and perhaps with other features as well, rendering it more suitable as an abode for life? When that supernova lit up the Galaxy, was it a primordial Star of Bethlehem, which announced not the salvation of mankind but rather its ultimate emergence?

Interstellar matter may gather in patchy clouds. The formation and collapse of such clouds is aided by the action of the spiral arms. (Courtesy Hale Observatories)

A supernova, exploding not far from the nebula which gave rise to our Sun, may have intruded tongues of gas into the nebula, as in this photo. (Courtesy Hale Observatories)

Yet this cannot truly be so. If it took much longer for stars to evolve to supernovahood than for a collapsing gas cloud to form its complete retinue of stars, it would be a rare matter of chance for a supernova to trigger star formation. By the time the largest of the new-forming stars had evolved and exploded, there would be no more left of the placental cloud from which more stars might form. But in fact the collapse of a cloud is far from a quick or uniform event; large parts of it may resist collapse, or else compress only slowly over millions of years. There is ample opportunity for part of a cloud to collapse, form stars, give rise to a supernova, and thus trigger the collapse of still other regions of the cloud. However, stars do not necessarily stay in the environs where they were born, but often move long distances. By the time a bright star is ready to explode, it may be far away from the remnants of its cloud.

We do not know. We do not yet understand how important supernovae are in triggering the formation of stars like the Sun. Most solar-type stars form almost certainly without this intervention. Our own case has probably been repeated again and again to form worlds unknown. There may be some degree of uniqueness in our Solar System being fathered by a supernova, but how much so we cannot say. Yet how piquant it will be if it is found that virtually all sunlike stars arise quietly, gently, following passage of a galactic density wave, but that we have the special distinction of a system rendered particularly favorable to life from having been conceived in violence.

These earliest phases of the Solar System's history are shrouded in questions unanswered, but the

As a nebula collapses, it forms small dark blobs that contract further to form stars and solar systems. (Courtesy Hale Observatories)

A new-forming star develops a disk of rotating matter that contains the material from which planets may form. (Courtesy Don Dixon)

next stages in its formation are much better understood. At an early date, the matter of the Solar System gathered itself together under its gravity and separated itself from the rest of the protosolar cloud. It was somewhat as though this cloud were a glassful of water tossed out a window, which separated into drops; one of these drops was the solar nebula, the direct antecedent of Sun and planets. The solar nebula also proceeded to contract, to compress itself due to its gravity. As it did so, its rotation, which it had inherited from the Galaxy, speeded up, forcing its matter to spread out to form a disk, perhaps ten billion miles across. In form it resembled the Milky Way, but was a hundred million times smaller.

By no means was this disk a steady, stable structure like Saturn's rings. It was much more like the atmosphere of Jupiter—wildly turbulent, driven by raging storms such as no mariner ever faced. Strong currents resulting from variations in pressure and temperature stirred it vigorously. New nebular material, itself turbulent, continually fell upon it, agitating it even more. Parts of the disk contracted or condensed further under gravity, producing large gaseous blobs, which moved through the primitive solar nebula and produced still more turmoil.

This tumultuous activity produced movement resembling the cracking of whips. A whip is long, flexible, tapering; when we crack it, we start with a rapid motion at its thick end. This motion produces a wave, which moves along its length. As the wave advances, the decreasing thickness of the whip makes the wave speed up. By the time the wave approaches the end of the whip, the thinness of its tip has brought the wave to and beyond the speed of sound. The crack of a whip then is actually a sonic boom produced by the supersonic speed of its tip, the first man-made object to exceed the speed of sound.

In the solar nebula the turbulent motions produced waves that traveled toward the surfaces of the disk, where the gas was less dense. As these waves advanced into thinner gas, they sped up; and when they passed the speed of sound, they became shock waves. As these shocks penetrated toward the disk surfaces, they became so powerful that they blasted nebular material back into space. In this fashion the disk lost mass. It evolved from a state containing as much matter as the Sun to a disk that had only perhaps one-thirtieth that amount, or thirty times the mass of Jupiter.

In this condition there was, at last, a reasonable degree of calm. The nebular disk was now too rarefied to sustain its former storms and turbulence. Now, for the first time, solid grains of dust, droplets of ice, and particles of rock could form. These condensed within the nebula, just as if they were hailstones condensing within a cloud. And like hailstones, they rained downward—not into the growing protosun, but to the midplane of the disk. These particles formed a thin disk within a disk, concentrated midway between the upper and lower surfaces of the gaseous nebula. The solar nebula thus resembled a hamburger sandwiched between two buns, or a phonograph record midway between two Frisbees.

This dust disk was the direct antecedent of Earth and the planets. Once it had formed, it behaved somewhat like a thin film of water, which beads up into many small droplets due to surface tension. The dust disk had no surface tension, but it did have gravity. It broke up into numerous small regions, and in each of these regions gravity caused the dusty material to coalesce. In only a few thousand years, the rocky or icy material of the Solar System was gathered into small bodies and planetesimals, which in the vicinity of what would be Earth were a few miles in diameter.

The planetesimals in turn bumped and jostled together as they went along, each to its separate orbit. Where they collided at high speed—a hundred meters per second or more—they tended to fragment, to shatter one another back to the primordial dust from which they formed. But at lower collision speeds, they tended to stick together. In this fashion many of the planetesimals gathered together, forming larger bodies. Eventually, in each of several regions of the Solar System, one such body grew larger than the rest. The transition from planetesimals to planets now was unstoppable.

At some time during these events, Jupiter and Saturn formed. They may have begun as rocky or icy cores, larger than others, which attracted much gas from the surrounding nebula because of their gravity, or as condensations or accumulations of gas within the nebula, which attracted many planetesimals from surrounding space. We do know that they formed quickly, during a time of no more than a few million years, and before the Sun itself was well formed. For in all of this, there was as yet no true sun.

In the inner regions of the disk, gases collected to form a massive concentration. As this concentration developed, it lit up with the first solar fires. These were not the true thermonuclear fires of stars, but rather the heat produced as the protosun compressed itself under gravity. As it compressed, it generated very strong turbulence in the inner regions. As in the cracking of a whip, this turbulence again sent shock waves coursing through the outermost layers of the protosun, blasting great quantities of material outward. In this fashion there flowed outward a steady stream of gases, at the rate of several million trillion tons per year. This stream scoured the Solar System, cleansing it of dust and gas, and removing what was left of the solar nebula. Jupiter and Saturn now would grow no more.

After some tens of millions of years of this so-called T Tauri phase, the Sun's interior finally reached the temperatures and pressures at which thermonuclear reactions could begin. The Sun now

atoned for its turbulent youth by beginning a ten-billion-year career as a steady, reliable source of

Within the solar nebula dusty material concentrated to form planetesimals, which in turn accumulated to form planets. (Courtesy Don Dixon)

energy. In the newly cleared Solar System, warmed now by this new star, the final accumulations of planetesimals and protoplanets were swept up by growing planets or scattered by Jupiter to yield the last stages of planet growth. This was a final flurry of violent action, as the impacting bodies tore great craters and basins in the young planet surfaces. Then even this subsided. The violent energies were spent; the clashing upheavals had run their course. There was peace among the planets.

Astronomers disagree endlessly over the details, but many would agree that the origin of the Solar System took place more or less as described. Then, is there anything special or unique in this? Did our Earth sidestep some pitfall that must have befallen many other forming planets, so that our Sun gained a retinue of planets where it might well have had only a scattering of dust? The answer is yes. There is good reason to believe that rocky planets like Earth are far from usual, that there is a "planetary hang-up," which causes most stars to form without such companions.

This hang-up begins in the stage when the solar nebula first forms a disk. Such disks are known to be unstable and tend to form the shape of a bar. As the nebula further evolves, the bar can divide in two, forming a binary star. Such binaries are indeed quite common, and can take various forms. If one of the companions has more than one-sixteenth the mass of the Sun, it will shine like any star.

If its mass is between one-sixteenth and one-hundredth the Sun's, though, it will not succeed in lighting internal nuclear fires but will glow with the feeble heat of gravitational compression. This   **25**

glow continues for a billion years, then ceases; and the star thereafter persists as a "black dwarf,"* a massive body which emits no light. If it has less than one-hundredth of solar mass, it already is no more than ten times the mass of Jupiter. Such a body glows only briefly, if at all, and we would call it a planet.

In 1976 the astronomers Helmut Abt and Saul Levy used the 84-inch telescope at Kitt Peak, Arizona, in a careful study of stars similar to the Sun. They examined 123 such stars, all within 85 light-years of the Sun, and all visible to the unaided eye. They were not the first to examine these stars, but they had better equipment than earlier astronomers, so their results were more complete. They concluded that of these 123 stars, 83 have binary companions that are true stars. They were unable to get good data on the existence of black dwarfs and planets, but they suggested there would be 20 black-dwarf companions, and 25 companions that would be planets resembling Jupiter. In other words, it appeared likely that all 123 sunlike stars have companions, ranging from Jupiter-size to sizes similar to the Sun itself.

What difference does that make? The answer is that such companions tend to prevent the growth of planets such as Earth. If a star has a binary companion (and most do), then it will be most doubtful that small rocky planets can form. This is the planetary hang-up, and we ourselves nearly fell victim to it. Astrologers speak of the influence of Jupiter over the lives of mortals, but their speculations are mere trifles. The truth of the matter is far more chilling: Had Jupiter been only slightly larger or nearer, or on a slightly less regular orbit, Earth as we know it would not exist.

Recall those early days when the matter of Earth was bound up in billions of small planetesimals, which had to collide and stick together if proto-earth was to grow. To do this, it was essential that conditions be steady and even, that the planetesimals be kept on orbits that were very nearly circles. If their orbits departed from this, and became even slightly elliptical, the planetesimals would collide at speeds too high to stick together and instead would tend to shatter. As long as there was no external disturbance on the planetesimal orbits, they could maintain their circular orbits, and the growth of planets would proceed apace. However, only a few hundred million miles away was Jupiter.

Jupiter's orbit was elliptical, and Jupiter, pulling with its gravity on the planetesimals, tended to shift these small bodies onto orbits that were also elliptical. This effect was so strong that had there been no countervailing effects tending to reduce Jupiter's influence, Jupiter would have prevented planetesimals from combining anywhere in the Solar System.

Fortunately, the presence of the solar nebula tended to soften or attenuate Jupiter's influence. Each planetesimal responded to the gravity not only of Jupiter but also of the nebula itself, and to the extent that Jupiter had only a small fraction of the nebula's mass, Jupiter's effect was reduced to a small fraction of what it would otherwise have been. Since the nebula took time to dissipate, the planetesimals were granted a reprieve. As long as the nebula existed, Jupiter's disruptions were held at bay. Still, the nebula's presence was far from a foolproof security. When the nebula mass fell below about ten times Jupiter's mass, its attenuations became insufficient to keep Jupiter from disrupting growth in a region outward from about three times Earth's distance from the Sun. Some time later the nebula lost more mass and had no more mass than Jupiter; then the disruption of growth crept inward as far as Mars. Not long after, it was the turn of Earth to feel the disruptions; but by then, fortunately for us, there had been enough growth to permit Earth to form in spite of Jupiter. The nebular reprieve had lasted long enough.

---

*Not to be confused with a black hole, which is hundreds of times more massive.

the T Tauri phase, a star becomes violently active and blows off much gas and dust into interstellar space. This star is doing ...t but around its equator a dense nebular disk prevents material from escaping. In time the disk will be blown away, leaving ...hind the planets which grew within that disk. (Lick Observatory photo courtesy George Herbig)

In most binary systems the action of the double star prevents planets from forming by accumulation of planetesimals. However, if the binary companions are very close together, planets still may form. (Courtesy Don Dixon)

Still, Jupiter was not through yet. In its very act of dissipating or reducing its mass, the nebula, which heretofore had acted to weaken Jupiter's influence, now introduced a new effect. It cooperated with Jupiter, and in so doing, produced a new threat to the growth of planets.

To understand this effect, we must think of an orbit as an ellipse, a flattened shape similar to an oval, which points in some definite direction. We may recall the game of Spin the Bottle, and imagine the bottle is ellipse-shaped. The direction to which an orbit points is not fixed, but slowly changes as the orbit shifts; we call this shift a precession. All orbits precess, usually at different rates, due to the various gravitational tugs they experience from different planets. Ordinarily precession causes no problems.

If a planet should find its orbit precessing at the same rate as Jupiter, however, something new will happen. The planet's orbit, even if originally a perfect circle, will become elliptical. It will become more and more noticeably so for as long as its precession is matched to Jupiter. If this lock is not broken, the orbit may become so strongly elliptical that the planet will cross the orbits of other planets, so that eventually it would be destroyed in a collision of worlds.

What controls this precession? For Jupiter and Saturn, it is mostly the gravity of one another and, in early days, that of the nebula. For a planet, or for the planetesimals from which it might form, the precession is controlled mostly by Jupiter, Saturn, and the nebula. The importance of the nebula depended on its mass. As this mass diminished, at any distance from the Sun there would be some special value for the nebula mass that would make the precession rates of local orbits match that of

Jupiter. What would happen next would depend on how fast the nebular mass was changing. If the change was slow, the precession rate would stay close to Jupiter's for a long time, and an orbit would become elliptical indeed. Only if the change was fast could orbits stay nearly circular.

Now let us venture inward from Jupiter's orbit, which is 483 million miles from the Sun, and examine the damage done by these effects. For the first 150 million miles or so we find virtually nothing more than empty space. There may once have been bodies orbiting here, but they apparently were swept up or ejected by Jupiter. This fact is ominous, for only if their orbits were markedly elliptical could they have suffered such a fate.

From about 320 to 200 million miles from the Sun, we find a collection of orbiting bodies, the asteroids. Here are no large planets or regular, nearly circular orbits. Instead, here are thousands of small objects, none of which are more than a few hundred miles across; most are much smaller. Their orbits are very elliptical, several times more so than Jupiter's. Even more telling, the total mass of asteroids is small; some one ten-thousandth that of Earth.

Here amid the asteroids we see the wreckage, the devastation wrought by Jupiter. For a brief halcyon time, the nebula was massive enough to keep Jupiter at bay so that planetesimals could begin to combine and grow. The early dissipation of the nebula allowed Jupiter to interrupt this growth. Then as the nebula slowly continued to dissipate, the precession effect moved with savage force amid these small, weak bodies, wrenching their orbits into ellipses. With this, the asteroids collided full tilt, at speeds of several miles per second, returning many of them to dust. This grinding-down or shattering continued till there were simply too few asteroids left to collide very often. Of the rest of them, there was nothing left but dust blowing in the solar wind.

We leave these scenes of carnage and destruction and proceed farther sunward. At 140 million miles from the Sun is a planet, Mars. Yet is it really a planet, or merely the largest of the asteroids? It is so small and shrunken, only a tenth the mass of Earth, and its orbit still is much more elliptical than Jupiter's. Evidently the planet-forming processes went further here, but in the end again it was Jupiter that carried the day.

Finally, 93 million miles out, we do find a true planet—large, fully formed, on a nearly circular orbit. Evidently, in Earth we have at last reached beyond the effects of Jupiter, have entered a region where the king of the planets ceases to reign. But. . . .

We are already four-fifths of the way from Jupiter to the Sun!

We look back over our shoulder at Jupiter and shudder. It was so close, such a near thing. These disruptive effects, which penetrated with full force so far sunward, depended only on Jupiter's mass and orbit and on the history of the nebula. When compared to other binary-star companions, Jupiter is so small, its orbit so much more nearly circular than most others. Had Jupiter been slightly larger, nearer to the Sun, or more elliptical in its orbit, then Mars today might be merely a scattering of asteroids and Earth a small, shrunken, wizened planet like Mars. We may think of the mythology of the Greeks, and be thankful that mighty Jupiter did not hurl its gravitational thunderbolts far enough to harm our planet. It was a very near thing.

Yet while we might search star after star and find nothing but binary companions and scatterings of asteroids, still there is hope. It is no accident that the orbits of Jupiter and Saturn are as nearly circular as they are, for these planets grew by accumulating gas from the nebula. As they did so, their orbits became more circular.

Moreover, many stars form not as binaries but as members of triple or multiple systems. These stars then follow strange, unusual orbits, which may result in the ejection of the smallest of them.

They are literally thrown out due to close encounters with the gravity of their fellows. In just this fashion were the spacecraft *Pioneer 10* and *11* ejected, following close approaches to Jupiter that gave them enough energy to escape. If a star is ejected early enough, it will still have the form of a solar nebula, and its evolution may proceed free of disruption from a binary companion or even perhaps of a large planet or black dwarf.

There is a third possibility. A king may wreak havoc with his army, but in his own castle there is calm. Similarly, Jupiter, disrupter of worlds, has in its immediate vicinity a most regular and stable collection of large satellites, which could well be regarded as planets in their own right. There may be many such systems of worlds, dancing attendance upon the binary companions of sunlike stars. It may be that what the binary companions take away via the planetary hang-up, they then return by forming planetlike bodies in their immediate neighborhoods.

Even so, the existence of the planetary hang-up gives us reason to appreciate Earth's uniqueness. We thus may continue onward in meditating upon our significance in the Galaxy; we may look farther, probe more deeply, and in particular we may consider the origins of life. And as we do so, we may remember the words of Paul of Tarsus: "We know in part, and we prophesy in part. . . . Now we see through a glass, darkly."

Even if the only planets that form are gas giants like Jupiter, they still may have systems of large satellites which resemble planets. This Voyager photo of Jupiter shows its satellites Io and Europa. (Courtesy Jet Propulsion Laboratory)

CHAPTER 2

The Terrestrial Hang-up

In thinking about the origins of life, it is hard to avoid the temptation to add the caveat, ''as we know it.'' On Earth, living organisms are built from compounds of carbon and nitrogen and live with the aid of oxygen. As Carl Sagan has aptly noted, some scientists claim that life must inevitably be so, but these scientists are biased since they themselves are built from compounds of carbon and nitrogen and breathe oxygen. Is this bias reasonable?

One alternative is that life might be based not on carbon but on silicon, which is chemically quite similar and in addition is more abundant. A great advantage of carbon for life is that when it bonds with other atoms to form compounds, the bonds are relatively weak and easily broken. Thus such compounds can partake in very subtle reactions. Also, carbon forms an incredible variety of compounds and combines readily with such key elements as oxygen, hydrogen, phosphorus, nitrogen, and sulfur. Silicon, by contrast, forms very tight chemical bonds; rocks are compounds of silicon, which is why they are hard and not susceptible to change. Still, a variant group of compounds known as silicones indeed are very similar to some carbon compounds. Could silicones serve as a basis for life?

To answer this, we must think of the first origins of life. Life began amid mixtures of simple compounds: ammonia, methane, water, hydrogen cyanide, carbon dioxide, phosphates. These were stirred and combined by the action of various natural energy sources, eventually forming more complex compounds: amino acids, the precursors of proteins; nucleic acids, the incipient genetic code; simple sugars; fatty acids; and the like. Charles Darwin had the basic idea as early as 1871, when he wrote: ''If (and oh! what a big if!) we could conceive in some warm little pond, with all sorts of ammonia and phosphoric salts, light, heat, electricity, etc., present, that a proteine compound was chemically formed ready to undergo still more complex changes, at the present day such matter would be instantly devoured.''

How might we introduce silicon into such reactions? Carbon dioxide is CO_2; the analogous silicon compound is SiO_2. But this is simply sand, a very stable compound not readily incorporated into chemical reactions. We begin to get somewhere by thinking of methane, CH_4; its silicon counterpart is silane, SiH_4. This compound is not at all so tightly bound and might well take part in biochemistry. The trouble is, it is like the Wicked Witch of the West in *The Wizard of Oz*. It is very sensitive to water. Even small traces of water will break it apart, and water is very, very abundant. Water (or steam) then would act to poison the prospects for silicon-based life by blocking at the start the pathways by which it could arise. (This does not rule out the possibility that silicon-based life may arise out of carbon-based life, but this idea will be deferred to another chapter.)

Somewhat more sanguine hopes may exist for another often-suggested possibility: life based not on water but on ammonia. Ammonia in many ways behaves similarly to water and is liquid over the range of temperatures of $-108°$ to $-28°$ F. This is rather narrower than the range for water, $32°$ to $212°$ F, but this need not prove to be a handicap. There are many small, cool stars around which worlds might orbit, with low temperatures suitable for lakes or oceans of ammonia.

The biochemist P.M. Molton has carefully examined how ammonia could substitute for water in the chemistry of life. This substitution would have to cover a number of important processes: action as a solvent; maintenance of a chemical balance within cells; formation of compounds analogous to proteins, fatty acids, nucleic acids, carbohydrates, lipids, steroids, phosphate compounds; production of energy, by processes akin to respiration. In a beautiful and detailed paper, Molton showed in 1974 how ammonia could do all those things and more. He even found ammonia-based analogues to the main energy-producing reactions in cells, the so-called Krebs cycle. In the Krebs cycle, citric acid serves to speed the breakup of glucose (a form of sugar), which is combined with oxygen to yield energy. If ammonia-based creatures exist, they would not breathe oxygen and exhale CO_2. Instead, they might well breathe (or drink) ammonia, and excrete a substance known as cyanamide, NH_2CN.

However, there is a problem in this. Such a planet would be bathed in ultraviolet radiation from its parent star, which would penetrate through the atmosphere to ground level as well as many feet deep into the oceans of ammonia. The radiation would pose a serious danger to life since it breaks proteins apart. On Earth we are shielded from this by a layer of ozone high in the stratosphere; the small amount of ultraviolet that leaks through nevertheless is quite sufficient to burn the skin, as a visit to the beach on a sunny day will quickly prove. In the absence of ozone, a form of oxygen, small living creatures could not survive on land. In fact, an ammonia-covered world would absorb oxygen, preventing its buildup in the atmosphere. There is no reason to think that ammonia-based proteins would be any less susceptible to damage from this radiation than are Earth's water-based proteins. Of course, we can speculate about the ammonia creatures sheathing themselves in shells resembling the ultraviolet-blocking windows of airliners; but on the basis of Earth's geologic record, we find this does not happen. Until our ozone layer formed, some half a billion years ago, Earth's life was effectively restricted to simple, unintelligent forms living in the sea.

Since we are interested in the origin of beings ultimately capable of colonizing the Galaxy, it is indeed quite reasonable to restrict attention to life as we know it, based upon carbon and water. Then, how did we come to be? And how readily might the same evolution have occurred elsewhere?

It is a commonplace idea that Earth had to form at the right distance from the Sun, or else it would have been too hot or too cold for life. This is a much more complex matter than was realized even a few years ago, because the Sun has not always been at the same brightness that we see today. It has grown brighter over geologic time, emitting more energy, while Earth has stayed at the same average distance from it.

This brightening of the Sun is a very well-supported finding in astrophysics. It results from the buildup of helium in the Sun's core, produced by fusion of its hydrogen. The exact amount of this brightening is somewhat uncertain, but the Sun today is probably about 35 percent brighter than it was 4.5 billion years ago, when Earth formed. (Some investigators have found values as high as 50 percent.) This means that if Earth existed then as it does today, it would have been colder, since the Sun was dimmer—so much that the oceans would have frozen solid. The subsequent slow brightening of the Sun would not have melted these oceans, since an ice-covered Earth would reflect most of the sunlight back into space. On such a world, life as we know it could not arise.

Evidently Earth escaped this fate. But how? The answer is that Earth's primordial atmosphere was very different from what it is today. It functioned as a blanket that trapped heat, so that primitive Earth was warm enough to keep the oceans from freezing. This blanketing is known as the "greenhouse effect," because Earth's primitive atmosphere acted like the glass windows in a greenhouse, which allow sunlight to enter freely yet trap the resulting heat. Even today the greenhouse effect warms Earth's surface by some 60° F. Several billion years ago this added warming may have amounted to over 170° due to the presence of ammonia and methane, which produce a large greenhouse.

So in order for Earth to avoid being too cold for life as the Sun was warming through geologic time, at all times there had to be enough greenhouse to keep the oceans from freezing. According to recent studies, this means maintaining an average temperature of at least 41° over the entire planet.

Why would a smaller average temperature freeze the oceans? The reason is that since this is an average, the polar regions would be much cooler. They would form massive ice caps, reflecting back more sunlight, reducing further the polar temperatures and causing the ice caps to grow. As this process continued, glaciers would creep toward the equator, and the polar ice regions would spread as the process fed on itself. Whether the glaciers would actually reach the equator would depend on Earth's initial average temperature, since obviously the glaciers would slow their advance and tend to come into a balance with unfrozen, warmer equatorial regions. But below this initial average of 41°, there would be no such balance. This runaway glaciation would freeze the entire planet.

There also is the matter of Earth being too hot for life. This does not mean being close to a hellishly hot sun, which beats down with an unforgiving intensity that would boil water. Instead, it is a matter of having too much of that good thing, the greenhouse effect. There is such a thing as a runaway greenhouse, and that is what happened to Venus.

Venus today has temperatures of 800° to 900°, but early in its history it may have been as pleasant and clement as Earth. Initially, Venus's atmosphere contained little or no water vapor or carbon dioxide, but active volcanoes soon released these gases in copious amounts. Both these gases trap heat effectively and have a substantial greenhouse effect; thus their release caused Venus's temperature to rise. As more of these gases accumulated, the Venus greenhouse grew hotter, and water was never able to condense to form oceans. The importance of oceans is that they dissolve the CO_2, which then can combine with calcium to form limestone. Earth's massive deposits of limestone contain the CO_2 that on Venus produced its thick, dense, smotheringly hot atmosphere. This was Venus's runaway greenhouse: a rise in surface temperature due to buildup of CO_2 in the atmosphere, which was not counterbalanced by other effects.

Someday Earth will suffer a similar fate. In aeons to come the Sun will continue to evolve and grow brighter, and there will be a warming of our oceans. With this, some of Earth's limestone will dissolve, releasing more CO_2 into the atmosphere, and more water will evaporate. The added CO_2 and water vapor will enhance Earth's greenhouse, producing still more temperature rise, more CO_2, more water vapor. All the while, Earth will be radiating heat back to space, and will tend to strike another balance, preventing temperatures from getting still hotter. This balance exists today, limiting Earth's greenhouse to the 60° rise mentioned, but the balance will shift to higher and higher temperatures as the sun slowly brightens. Eventually the balance point will be hot enough to allow the equatorial oceans to boil. With this, all hope will vanish; in a short time all Earth's water, and much CO_2 from limestone, will enter the atmosphere. An interplanetary explorer will find Earth and Venus all but identical.

This digression into planetary science, this discussion of the runaway greenhouse and of runaway glaciation, sets the stage for a short history of Planet Earth. We may think of our planet as a comforting mother, but her history in fact was much more like the Perils of Pauline.*

Four and a half billion years ago the newly formed Earth was a ball of rock and iron, laced with radioactive potassium and uranium, whose radiant decay heated the outer layers of Earth, producing volcanoes on a scale that can only be called Promethean. Amid tempestuous rumblings and along massive fissures or fault lines the whole world over, there were cone-shaped forms of young erupting volcanoes. Now smoking, now roaring, now merely glowing balefully with lava in the throat, time and again they expelled the red-hot magma, or exploded in sheets of fire. As the molten rock poured forth, great clouds of gas issued from vents and shrouded the young planet in mist. The volcanic gases were nearly all water vapor and CO_2, but about 1 percent of their content was methane, and a somewhat smaller fraction was ammonia. Despite their small quantities, these last two gases provided most of the early greenhouse, so that Earth would not freeze in the feeble warmth of a younger sun.

The water vapor gathered in vast extents of clouds, then condensed in torrents of rain, which gathered to form the incipient oceans. Part of the CO_2 entered the atmosphere, but more and more of it formed limestone as the oceans grew. Much of the ammonia broke down to form our atmosphere's first nitrogen, but some remained. With the growing content of methane, these gases were Earth's first atmosphere.

Never again would conditions be so favorable to the origin of life. Beneath the atmospheric greenhouse, mild, equable temperatures prevailed. The early shallow seas were stirred by volcanic eruptions and bombarded with solar ultraviolet radiation, lightning strokes, and the shock waves of impacting meteorites. These created chemical reactions in the atmospheric gases, producing organic chemicals of increasing complexity. In this, the ammonia and methane did more than produce a greenhouse, more than provide raw materials for the reactions of life. They prevented oxygen from destroying the delicate biochemicals. High in the atmosphere, solar radiation was breaking water vapor molecules and releasing oxygen, a deadly poison that would have destroyed the primitive carbon compounds evolving toward life. But as soon as any oxygen formed, it immediately combined with the ammonia or methane and was removed.

---

*The computations for this history were made by Michael H. Hart of the Laboratory for Planetary Atmospheres, NASA-Goddard Space Flight Center.

The formation of Earth's first atmosphere took place amid the violent scenes that marked the final stages of its origin. (Cour *Don Dixon)*

Erupting volcanoes and fire fountains yielded the rocks of Earth's crust, and also released the copious water vapor and carbon dioxide of the primitive atmosphere. Methane and ammonia also were released. (Courtesy Don Dixon)

The precious biochemicals formed, polymerized, isolated themselves in droplets, and slowly groped toward life. This was chemical evolution, or survival of the fittest compounds. It was not yet the true evolution of Darwin, but it was to the degree that chemical structures formed with more and more of life's attributes and tended to persist while others were destroyed and broken back to simpler chemicals. Within a few hundred million years of Earth's origin, true life appeared, able to consume food and grow and to reproduce with a genetic code.

This early life was in the form of one-celled creatures resembling bacteria or algae. They had an extremely simple structure and were very small in comparison to present-day cells; yet how promising they were as they drifted to and fro, warmed by sunlight in the estuaries of ancient seas. They lived by consuming the complex biochemicals that had not evolved as far as life. By simple chemical steps resembling fermentation, these cells could break down their foods' structures to release a modicum of energy. These processes were wasteful, inefficient, and utterly dependent on nature to synthesize more of these biochemicals.

The challenge was for these cells to develop better means of getting energy. The evolving bacteria responded by developing enzymes and organic dyes, which could obtain energy from the iron and sulfur compounds dissolved in the sea. With further evolution, there arose blue-green dyes related to chlorophyll. In some long-forgotten lake, the first verdant traces of cyanobacteria or blue-green algae appeared. Henceforth there would be plant life to grace Earth's waters.

For all this, the prospects for life still were tenuous. As volcanoes continued their eruptions, the content of atmospheric methane and ammonia grew greater, its greenhouse stronger. Lacking this effect, Earth's average temperature would have been $-65°$ or worse; with it, the average was 106°

four billion years ago, and heading higher. There now was certainly no danger of runaway glaciation; quite the contrary. The danger now was from the runaway greenhouse. If the average were to reach 126°, this would become unavoidable, and life would perish in steam.

As Earth approached its billionth birthday, the volcanoes began to decline in power and the rise in average temperature became less steep. Yet the rise continued to 110° and even higher. Now, however, the growing presence of life began to make itself felt. As the algae spread, they carried out photosynthesis and produced oxygen. The oxygen combined with the ammonia and methane to remove some of these gases, and the rise in their content slowed as the work of the algae approached and then matched the work of the volcanoes.

For a hundred million years matters hung in the balance. Here were the volcanoes, diminishing in action yet still powerful, every eruption a reminder of Earth's peril. There were the algae, growing in thin mats, freshening the waters with the oxygen of hope. Here were globe-embracing thicknesses of cloud, fed by the plutonic fires, dimming the weak sunlight and bidding to make Earth permanently a place of fire and darkness, of stifling, choking, never-breaking overcast. There the algae were accepting the feeble light, growing and budding in their turn, taking up the challenge: Would life predominate, and remake Earth after its own nature? Or would fire and steam be the victors, and thereafter rule the world?

At last the issue came to a resolution, and we are evidence that the decision was on the side of life. The content of ammonia and methane peaked, then began to decline as the algae made their

The first oceans formed amid basins of naked rock, only recently erupted. (Courtesy Don Dixon)

presence felt with increasing strength. The greenhouse also peaked, with Earth's average temperature topping out at 111° or 112° F. Thereafter, temperatures began to decline. Earth had passed its first crisis in the evolution of life—and it was the humble blue-green algae that saved the world for us all.

The next billion or so years were increasingly peaceful. It was the time of simple one-celled life, now developed into a community of plants and animals; that is, of algae, which could photosynthesize, and of bacteria, which could engulf and consume the primitive plants. This was the first food chain, but life was still crude, primitive. Nor was it widespread. Since there was still no free oxygen in the sea or atmosphere, the cells lived by the inefficient process of fermentation. Most of the energy in food went to waste, for want of oxygen with which to extract it.

Life was limited in extent for another reason. Earth was bathed in solar ultraviolet, which rendered the land unfit for life and which penetrated deep into the oceans. No life could exist in the surface layers; yet algae could not grow too deep, for they required the pale sunlight that filtered through to depths where the ultraviolet could not reach. At greater depths the sunlight itself could not penetrate. In this thin layer of water, between the extremes of too much ultraviolet and too little sunshine, life found its milieu.

Slowly, slowly the levels of methane and ammonia fell while those of nitrogen rose. With this, temperatures fell too: to 88° three billion years ago, to 75° a half-billion years later. By now the Sun was beginning to evolve and its core showed the first traces of helium. Its temperature rose

Within a billion years after its formation, the Earth may have narrowly escaped having its oceans boil. Had this happened, Earth today would resemble the searing inferno which is Venus. (Art by Don Dixon courtesy NASA)

Some two billion years ago Earth may have experienced a time of great cold which sent glaciers marching toward the Equator and threatened to engulf the world in ice. (Courtesy Don Dixon)

imperceptibly, as did its brightness. Some two billion years ago, the oxygen from photosynthesis succeeded in removing the last traces of ammonia and methane; Earth's atmosphere then was nearly all of nitrogen, with a few percent of water vapor and carbon dioxide. Now a new problem arose. With these last traces gone, their greenhouse was gone too, and in the short time of a quarter billion years, temperatures plunged over twenty degrees. Now the danger of runaway glaciation became real, as the global temperature dropped into the forties. With nothing left but CO_2 and water vapor to give the atmosphere its greenhouse, this global mean fell to 46° two billion years ago; then 45°, 44°, 43°. The margin against disaster was only two degrees.

The ancient Norse saga, the Eddas, tells of the legendary Fimbul-winter, when the world was locked in ice and snow, with fierce and bitter winds howling from the dark and brooding cloudbanks to the north. Not for a mere hundred million years but for a billion, this was the world north and south of the tropics. Here was the majesty and splendor of the arctic, embracing far more than today's mere polar patches. Here was the Sun sparkling on the crystals of new-fallen fields of snow on rare days when the sky turned a brilliant blue. Then again there were the endless, trackless stretches of pack-ice, of jagged blocky floes extending past the horizon, past many horizons. There were glaciers, extending perhaps to lands where corals grow today, breaking at their edges to release massive bergs in sprays of splashing water and rime-ice. And always, always, the bitter whistling winds, in air thinner than

39

The early oceans were stirred by strong tides, for the Moon formerly was much closer to Earth than it is today. (Courtesy Don Dixon)

today's, their moanings and piercing calls ceaselessly repeating the message that those who live must die.

No life could survive in the lands of ice, but in the equatorial regions it made its stand. It could not now act to save itself, to alter its fate by growing and spreading, or by releasing oxygen. Still, the ocean and atmosphere were in equilibrium and there would be no reduction in the atmospheric CO_2 or water vapor to eliminate the slim margin of greenhouse effect that held back the glaciers. In the end, temperatures bottomed out just below 43° and slowly rose, degree by degree, as the billion years elapsed. It was not life that did this, or any other process of Earth. It was the Sun, slowly gaining its strength and brightness. As the aeons passed, the global mean temperature rose into the high 40s, then 50s. The danger of a world of ice was even more real than the early danger of a world of steam; but it too passed.

The low temperatures at the minimum had not threatened life, but as algae produced more oxygen, there came a time when that gas could begin to build up in the atmosphere. When the last ammonia and methane were gone, free oxygen appeared for the first time, in the amount of some 5 percent of the present level. We would find this quite inadequate; we would suffocate in such a world. Yet even this modest level of oxygen wrought a revolution in the history of life.

Instead of using fermentation, it now became possible for cells to obtain energy by respiration, by combining sugars and other foodstuffs with this oxygen, bringing an immense gain in the energy available to life. With fermentation, in the absence of oxygen, a molecule of glucose produced two units of energy; that is, two phosphate bonds in the substance known as adenosine triphosphate or ATP, which is the universal medium of exchange of living energy. With respiration, a molecule of glucose yielded not two but thirty-six such phosphate bonds. The consequence was somewhat as if a car's gasoline mileage were to improve eighteen-fold.

This change stimulated an increase in the complexity of life. The earlier cells had been small, with little or no cellular structure or internal parts. They were merely blobs of living matter enclosed within simple cell walls with their genetic material, DNA (deoxyribonucleic acid), as a loop within. The more complex cells which evolved were much larger, and came to have many of the features of cells in today's plants and animals. Their chlorophyll was bound up in specialized centers or chloroplasts. Other cellular structures, the mitochondria, arose to handle the digestion of food by respiration. Some of these cells developed efficient means of moving about with whiplike appendages known as flagellae. Most importantly, the DNA came to be organized into chromosomes located within cell nuclei. The evolution of nuclei and chromosomes brought modern forms of cell division and opened the way for the advent of sexual reproduction, which relied on these modern forms. This advance was a vast improvement over the primitive buddings and splittings of bacteria that had existed before.

In shallow seas, lying beyond the margins of the emerging continents, life first formed and found its milieu. (Courtesy David Egge)

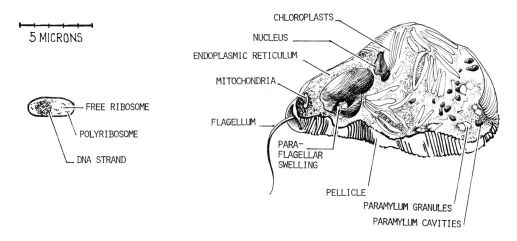

5 MICRONS

FREE RIBOSOME
POLYRIBOSOME
DNA STRAND

CHLOROPLASTS
NUCLEUS
ENDOPLASMIC RETICULUM
MITOCHONDRIA
FLAGELLUM
PARA-
FLAGELLAR
SWELLING
PELLICLE
PARAMYLUM GRANULES
PARAMYLUM CAVITIES

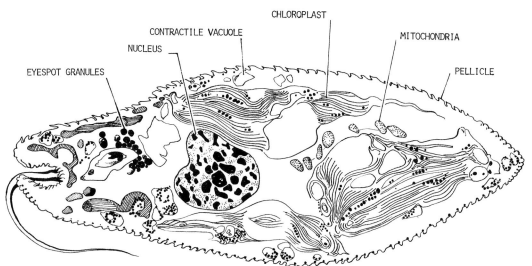

CHLOROPLAST
CONTRACTILE VACUOLE
NUCLEUS
MITOCHONDRIA
EYESPOT GRANULES
PELLICLE

The initial buildup of atmospheric oxygen permitted a vast increase in the complexity of single-cell life. The upper left-hand corner shows a simple bacterium, Escherichia coli, *typical of life before this oxygen buildup. The upper right and below show sectional and lengthwise views of* Euglena, *a complex single-cell life form that emerged after oxygen became present in the atmosphere. (Courtesy Gayle Westrate, Division of Biology, California Institute of Technology)*

The increase in cell complexity may seem minor, yet it was a greater step for life than the origin of plants and animals. As long as there were only the bacteriumlike forms of the days before free oxygen, there could be nothing more than one-celled life. The new, more complex cells could organize into tissues; life could come to be of many cells. That would not come about immediately, but even a billion and a half years ago, the revolution of more-complex cells had brought to these basic units of life the structure and functions that they have today.*

---

*What here are described as "bacteriumlike cells" and "complex cells" are known to the biochemist as prokaryotes and eukaryotes. All the cells in today's plants and animals are eukaryotic.

The initial buildup of atmospheric oxygen did not proceed very far despite the great impetus it gave to life. The solar ultraviolet still restricted life to regions beneath the sea; in addition, the oxygen soon found other chemicals with which to combine. Two billion years ago, the seas were full of dissolved iron. The initial free oxygen in seawater acted on this iron, and the seas slowly rusted. The dissolved iron changed from a soluble form to an insoluble one, and like snowflakes falling from the air, particles of iron oxide gently formed and fell to the seabed. These in time would be the world's economic reserves of iron ore, but that was far in the future. Even more oxygen was used up in reactions involving sulfur.

For a billion and a half years, oxygen levels in the atmosphere rose slowly, from 5 percent of present-day values two billion years ago to 8 percent. For much of this time life was still limited to the narrow region of water between the zones of too much ultraviolet and too little sunlight—between the devil and the deep blue sea, as it were. Nearly as fast as oxygen was produced, it combined with iron or sulfur. Still, life was not without its advantages. High in the stratosphere, sunlight transformed some oxygen into ozone, which shielded out some of the ultraviolet. This meant green plants could grow closer to the surface, and life could not only occupy a greater range of ocean depths but also could take advantage of the more intense sunlight in the shallows of oceans.

About half a billion years ago, with oxygen at 8 percent of the present level, the ozone layer grew thick enough to permit plants to grow in open air, without any protection by a depth of water. They could then grow on the very surface of the sea, taking full advantage of sunlight and releasing oxygen directly into the atmosphere. The protection of ozone meant that the first plants could creep landward from the shallows and colonize what heretofore had been bare rock. The result was a vast flowering or proliferation of life and a rapid rise in oxygen levels. A hundred million years saw the oxygen level quadruple to one-third the present level. By three hundred million years ago, the level had passed half the present value and was still rising.

This increase sparked a remarkable diversification of life forms. In the short time of the geologic epoch known as the Cambrian, when oxygen levels were quadrupling, virtually all the principal types of Earth's life forms appeared, including not only the major forms of invertebrates but also primitive creatures possessing a notochord, a structure which in modern embryos develops into vertebrae. This Cambrian revolution saw the advent of animals with bones or shells, which left permanent fossil impressions. The Cambrian epoch was followed by others but the pattern of life's development was clear: increasing complexity, increasing diversity. A few more hundreds of millions of years of Earth's history, and one of these complex forms would gain the wit to look back and contemplate these changes, and to wonder.

And among the things we may wonder about is: How much of this are we to believe?

There is good geological evidence for the first appearance of free oxygen some 2 billion years ago, for the first complex cells 1.5 billion years ago, and of course for the Cambrian epoch. Evidence for the early times of heat and cold is much more scanty and ambiguous. There is good evidence for the existence of liquid water some 3.2 billion years ago, but it does not help determine its temperature. Tillites, a form of rock associated with glaciers, have been found from 2 billion years ago, which would be expected if that was a time of great cold. Yet other scientists argue that Earth's climate then was even warmer than it is today. So it must be emphasized that the history of Earth given here has been derived from calculations and only partly from the geologic record.

However, a number of different scientists have studied the question of Earth's fate if it had formed slightly closer to the Sun, or slightly farther. These studies are much more pertinent to the true

The formation of an atmospheric ozone layer allowed the evolution of complex multicellular forms of life. These dinosaurs (genus Diplodocus) are surprised by the presence of water at the bottom of a familiar rift valley. In later aeons the valley will open wider and become the Atlantic Ocean. (Courtesy Don Davis)

concerns of this chapter and far less dependent on the geologic record. With minor variations, they agree with the conclusions to be drawn from the work of Michael Hart discussed here: Earth would have suffered a runaway greenhouse if it had formed more than 5 percent closer to the Sun. It would have frozen had it formed more than 1 percent or 2 percent farther out.

Thus did our Earth evolve, and the life upon it, according to these ideas. Many scientists have long argued that life would arise whenever conditions were favorable, but this does not necessarily mean the life would be more than the simplest bacteria or algae. There would have to be two other events before life could advance to offer the prospects of future intellect. First, there would be need for an initial quantity of oxygen free in the atmosphere to bring the revolution of increased cell complexity, producing cells as we know them today. Second, there would be need for increased oxygen levels to promote formation of the ozone layer and permit life to emerge from beneath the water. This event would lead to the Cambrian revolution, with its vast increases in plant and animal complexity.*

On Earth, these three revolutions—initial origin of life, origin of complex cells, and the Cambrian revolution—all came essentially to completion within a few hundred million years of their onset. Since this time is short compared to the billions of years of Earth's history, we are justified in

*These ideas were first put forth in a slightly different form by L.V. Berkner and L.C. Marshall in the mid-1960s.

saying that on other planets, similar atmospheric changes would soon bring about similar advances in life's evolution.

All these developments depended quite critically on three things, fulfillment of which was a matter of chance. The Sun had to be the right mass. The Earth had to be the right mass. And the Earth had to be the right distance from the Sun.

The last of these is the simplest to understand. Had Earth formed at less than 95 percent its present distance, the early greenhouse would have run away and our oceans would have boiled. From that day forward, Earth would have been a lifeless planet resembling Venus. Had Earth formed beyond 101 percent its present distance (that is, more than 1 percent farther out), it would have frozen over some two billion years ago due to runaway glaciation, in Hart's calculations, and today would resemble Mars.

What of the size of Earth? If it had formed a bit farther out so as to just avoid (by an even narrower margin) freezing over when the primordial ammonia and methane were removed, it could be no more than 30 percent more massive than it actually is. That is, its diameter could not exceed 8,639 miles, compared to the actual 7,926. A larger Earth would have had more internal radioactivity and hence would have formed its early atmosphere more rapidly, so much so that the early buildup of ammonia and methane, prior to the advent of photosynthesis in algae, would have taken place more quickly and forced a runaway greenhouse.

By contrast, if Earth had formed a bit farther inward so as to have avoided this initial runaway by a very narrow margin, it still would have had to avoid runaway glaciation when the algae removed the ammonia and methane. It thus could be no less than 14 percent less massive than it is, or with a diameter less than 7,530 miles. Otherwise, atmospheric evolution would have progressed more slowly; and when Earth faced the crisis of runaway glaciation, there would have been insufficient CO_2 and water vapor in the atmosphere to yield a greenhouse sufficient to avert this fate.

The matter of the Sun's mass also deserves note. If the Sun had formed with more than 10 percent additional mass, it would have begun emitting considerable amounts of ultraviolet radiation by the time it was four billion years old. This occurrence would have inhibited the Cambrian revolution. To combat this radiation, Earth would have had to have formed farther from the Sun. The presence of this excess radiation thus would have cut into the range of distances over which Earth could have formed so as to progress to the Cambrian revolution. If the Sun had formed with more than 20 percent additional mass, its evolution would have progressed so rapidly that by the age of four billion years its temperature would have increased very quickly. This would have driven Earth to a runaway greenhouse at virtually any reasonable distance.

A much smaller sun would have posed a different problem. Small stars evolve less quickly; the growth of their brightness with time is slower, less marked. Let us remember that 2 billion years ago, when Earth had escaped the early threat of a runaway greenhouse only to face runaway glaciation, what saved our world was that the Sun had increased slightly in brightness over the previous 2.5 billion years. With less than 83 percent of its present mass, the Sun would have evolved so slowly as to have failed to have offered those critical few degrees of extra temperature when needed. The removal of the last traces of ammonia and methane then would have brought runaway glaciation.*

The concept of Earth's history as being poised so delicately between fire and ice is reminiscent of Dr. Pangloss in Voltaire's *Candide,* who opined that everything is for the best in this best of all

---

*In these calculations, Michael Hart assumed that Earth could have formed at any distance from the Sun, so as best to improve its prospects for life.

There is a tendency for gas-giant planets to form rings. Where such planets exist in other solar systems, may they have earthlike worlds in orbit about them? (Courtesy David Egge)

possible worlds. Such a conclusion may seem strange, particularly at income tax time; yet it was Robert Frost in his poem, ''It Bids Pretty Fair,'' who put things in perspective:

> The play seems out for an almost indefinite run.
> Don't mind a little thing like the actors fighting.
> The only thing I worry about is the Sun.
> We'll be all right if nothing goes wrong with the lighting.

This then is the terrestrial hang-up: that to form a world suitable for life as we know it, there must be what actually is a very good replication, obtained purely by chance, of our system of Sun and Earth. So if ours is indeed the best of all possible worlds, more or less, then we should be able to estimate how many similar worlds may be found. This can be done by using the ideas of the planetary hang-up and the terrestrial hang-up.

To begin, the age of the Galaxy is known: some fourteen billion years. Throughout most, if not all of that time, stars have been forming, some twenty per year. However, not all of them would be of interest in a search for extrasolar civilizations. The first-formed stars were almost entirely of hydrogen and helium. As they evolved, they built up heavy elements such as carbon, oxygen, silicon, nitrogen—elements from which planets could form. The early history of element-building in the Galaxy is somewhat unclear, but from what we know of astrophysics it seems reasonable that for stars formed in the first three billion or so years, there were insufficient heavy elements to permit planets to form.

Also, stars that have formed relatively recently are too young for us to expect intelligent life to have arisen. Earth is a most flourishing habitat; there can be few planets on which life has developed more exuberantly. Yet it has taken all but the last few million years of earth's history to develop what we call, parochially, "intelligence." If our history is representative, then stars formed more recently than the Sun are not of interest. The age of the Sun is known: 4.5 billion years.

So we are interested in the stars formed between 4.5 and 11 billion years ago; these number some 130 billion. Not all of these are potential abodes for planets. Some are hot, bright, blue stars like Altair and Sirius. These burn out so quickly, in no more than a couple billion years, that their worlds will perish long before life can well develop. Many more stars are small and cool, the so-called red dwarfs. These will burn for hundreds of billions of years, but like a candle in a window, they provide little heat at a distance. Hence their planets would be expected to be too cold for life. To be more specific, we can construct a table:

PROBABILITIES FOR EXISTENCE OF EARTHLIKE PLANETS*

Stellar spectral type	Mean stellar mass (Sun = 1.0)	Fraction of stars with this spectral type	Habitable range of distances from star (Earth from Sun = 1.0)	Probability of finding a planet in that range
F6	1.18	0.00474	1.536–1.591	0.0621
F7	1.14	0.00484	1.382–1.445	0.0791
F8	1.10	0.00499	1.240–1.309	0.0965
F9	1.06	0.00512	1.115–1.181	0.1027
G0	1.02	0.00520	1.009–1.061	0.0894
G1	0.985	0.00531	0.921–0.961	0.0753
G2	0.955	0.00534	0.850–0.880	0.0612
G3	0.930	0.00586	0.793–0.815	0.0481
G4	0.900	0.00628	0.728–0.742	0.0334
G5	0.870	0.00751	0.666–0.673	0.0182
G6	0.850	0.00810	0.627–0.630	0.0083
G7	0.825	0.00892	———	0.0

*Data in the first three columns are taken from *Habitable Planets for Man* by Stephen H. Dole, (New York: Blaisdell, 1964), pp. 102 and 104. The fourth column was computed from equations given by Michael Hart; see references to his papers in the Bibliography. The last column again uses data by Dole, pp. 91 and 92.

Closeup view of Io, a large satellite of Jupiter. Such satellites appear to grow to the size of the Moon or Mercury, but there is no evidence that they grow large enough to resemble Earth. (Courtesy Jet Propulsion Laboratory)

A bit of explanation is in order. Astronomers classify stars by the appearance of their spectra, thus furnishing a convenient way to refer to groups of similar stars. The first column gives the astronomers' code for this classification. The fourth column gives the range of planetary distances over which Earth might have formed so as to avoid thermal runaways while progressing to the Cambrian revolution. The last column assumes that planet distances are distributed as in our own solar system. For example, a typical star in spectral class F9 has a mass 6 percent greater than the Sun; some one-half percent of all stars are of this type. One would find an earthlike planet only between distances from that star of 1.115 and 1.181 times Earth's distance from the Sun, but there is a bit better than a 10 percent chance that a given star of that type would actually have a planet there.

Of the 130 billion stars that are the right age, a great many of them suffered the planetary hang-up and failed to form worlds such as Earth. In view of the carnage wrought by Jupiter within our solar system, it is risky to imagine that planets exist where a star has a binary companion larger than

Jupiter. Following the estimates of Abt and Levy, discussed in Chapter 1, this would rule out 90 percent of these stars. We thus have 13 billion candidates.

From the table, we find that only about one-third of 1 percent of these would have the right combination of stellar mass together with a planet in the proper range of distances. We thus are down to 46 million planets. Earlier we noted that to avoid the end of life, a planet must be between 86 percent and 130 percent the mass of Earth. If planets throughout the Galaxy have the same distribution of masses as in our own solar system, then there is only a chance of 1.9 percent that a given planet has the right mass. This brings us finally down to 880,000 worlds like ours in the Galaxy.

We have not considered the suggestion that binary companions resembling large versions of Jupiter might have systems of worlds in close orbit, like the Galilean satellites. All four of our Solar System's large planets have such collections of large satellites, these being Jupiter, Saturn, Uranus, and Neptune. However, the mass of the largest Jupiter satellite, Callisto, is 0.0246 times that of Earth. Jupiter is 18.5 times more massive than Neptune, but the largest satellite of Neptune (Triton) is 0.0227 times Earth's mass or 92 percent that of Callisto. By coincidence, the mass of Triton is also very close to that of Titan, the largest satellite of Saturn. There thus is no reason to believe that even if a binary companion were much larger than Jupiter, it could possess a group of close-circling satellites one of which were Earth-size. The processes that produce such worlds seem to limit their sizes to roughly that of the Moon.

So we finally come down to an estimate of 880,000 planets suitable for life. Nevertheless, this now is the real McCoy, the genuine nitty-gritty. We expect that each of them will be virtually a dead ringer for Earth, of similar size and appearance, with a similar star for its sun, and at the right distance and age to have experienced the Cambrian revolution upward of half a billion years ago. Moreover, it is likely that every one of them is (or soon will be) the abode of an intelligent species. These are just the worlds we would seek in space. Yet by these calculations, there is only one such planet for every 227,000 stars.

Our world is as a speck of dust lost in vastness, but to describe our sun as an ordinary or average star, as in the astronomy texts, is not correct. It is a very special star; its uniqueness lies in its being one of perhaps only 880,000 that are abodes for advanced forms of life. So it is not difficult to imagine that the other barren stars are calling to us, inviting us to cross the vast gulfs of distance, to find new lands, to make them fruitful.

This is the dream of space travel. This is also the dream of space colonization. We may realize such dreams in time, if the call of the distant stars is not to be denied. Should we one day reach out in this way, our efforts will be seen as flowing from today's space projects, today's space hopes. It will be a task for the future's historians to trace the lineage from the space shuttle to the starship.

If this lineage is today unclear, still we may hope. It is with this hope that we may carefully examine our current space activities, with an eye to finding within them the germs of future advances. In this spirit, we may look carefully at the prospects for space flight along the way to the stars.

The stars are far, but close at hand is the space between Earth and the Moon. It is here that we may build. The possibilities in this near-space must necessarily concern us, and must be explored in detail, before we can speak again of the stars. These possibilities are the subject of the next several chapters, and the prospects to be kept in mind are that they will include, as an activity beneath the orb of the Moon, the colonization of space.

CHAPTER 3

Space Colonization Soon?

We may say that astronautics is the science of dreams, and in the early 1950s that science was already some fifty years old. The general public had long since been made aware of the existence of large rockets, such as the German V-2 of World War II, and looked to the development of even larger rockets for the then widely anticipated World War III. Rocket aircraft, flights to the Moon, exploration of Mars and of other planets—had all been widely discussed in the popular press. Science fiction was booming, and such movies as *Destination Moon* and *The Conquest of Space* were playing to large crowds. The public was primed.

So it was, in such an atmosphere, that the rocket scientist Wernher von Braun set forth a vision for the future. He proposed that the nation commit its energies and resources to a project that he said could be built with current technology: a space station. The cost was to be a mere $4 billion in 1952 dollars, some $15 billion in today's money.

Von Braun proposed to build and develop a fleet of huge three-stage reusable shuttle rockets, 265 feet tall and weighing 7,000 tons—more than twice the weight of the Saturn V moon rocket. Each rocket was to lift on the thrust of fifty-one rocket motors, generating a total thrust of 14,000 tons, compared with 3,750 tons for the Saturn V. The third stage would be a piloted, delta-wing craft, somewhat resembling today's space shuttle orbiter, carrying 36 tons of cargo to orbit 1,075 miles up.

The cargo would consist of elements for the space station proper, which would be a wheel-shaped affair, closely resembling the one made famous in the movie *2001: A Space Odyssey*. It would house two hundred crew members; even at that early date, Von Braun argued that the crew should include both sexes. The station was to be 250 feet across with three decks or levels and rotating to provide

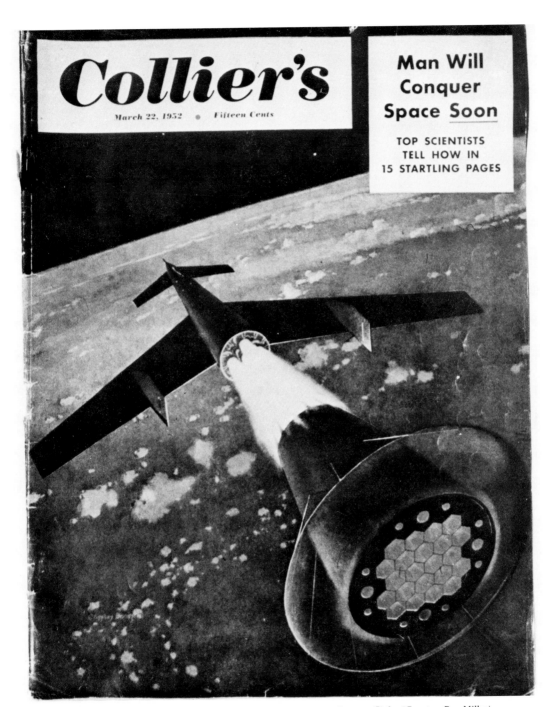

Collier's, *in its issue of March 22, 1952, touched off a wave of public interest in space flight. (Courtesy Ron Miller)*

In the Collier's *series, Wernher von Braun and other experts discussed the building of manned rockets which could fly to Mars. This three-stage rocket, shown in cutaway view, has fifty-one motors in the first stage. (Art by Rolf Klep, courtesy Ron Miller)*

one-third normal gravity. Its proponents suggested that such a station could carry atom-armed missiles and telescopes to peer down on the Kremlin, but as von Braun made clear in his book *The Mars Project,* the real purpose of such a space station would be to support a manned flight to Mars.

The feasibility of such a project can best be judged by examining the performance records of the day. The most advanced rocket airplane was the Douglas Skyrocket. In 1951 it touched 1,238 miles per hour at 79,000 feet—rather higher but a bit slower than the cruise conditions of today's commercial Concorde jets. The most advanced made-in-USA rocket was the Viking. When it worked properly, which was never a foregone conclusion, it would lift off with ten tons of thrust and reach an altitude of 135 miles. It did not inject a satellite into orbit at that height, but simply made an up-and-down flight like an arrow shot into the air. Eventually, with further development, Viking served as the first stage of Vanguard, which launched satellites into low orbits having the magnificent weight of fifty pounds.

Even today, a project such as von Braun's would represent no mean feat. In the 1950s it was widely publicized in magazines and TV shows and generated much attention among science-fiction fans. More serious rocket specialists were inclined to dismiss it as at best a dream for the distant future, at worst a hopeless fantasy. And yet it could not be ignored, for von Braun was the world's foremost and most experienced rocket designer.

This bit of history is worth remembering when contemplating the colonization of space. The idea of space colonies traces back to the earliest space pioneer, Konstantin Tsiolkovsky. Most of the important concepts can be found in the writings of Krafft Ehricke, Arthur C. Clarke, and particularly Dandridge Cole, whose 1964 book, *Islands in Space,* was entirely devoted to the topic. Since 1974, Gerard K. O'Neill, professor of physics at Princeton University, has led the modern studies of the subject. It is O'Neill who has organized the Princeton technical meetings beginning in 1974, as well as the study programs at NASA's Ames Research Center, which have brought space colonies to the serious attention of both the public and the aerospace industry.

The most significant of these studies took place in the summer of 1975 and provided much of the subject matter for my earlier book, *Colonies in Space.* The final study report was regarded as so important that it was published by the Government Printing Office as a hardcover book, with a foreword written by the head of NASA, James G. Fletcher. Although many details have been revised, the basic outline of space colonization presented there remains the accepted one to this day.

This view of a space colony called for an orbiting city, as far removed from an Apollo capsule or space shuttle flight deck as is a California seacoast town from a cave. The people would live in garden apartments with two or three bedrooms, surrounded by blooms of color. There would be flowers and lush gardens, parks and fruit trees, ivy and lawns—and, not to be outdone, the apartments themselves would sport stuccoed walls and tinted window panels.

Inside would be drapes, carpets, comfortable furniture, all imported from Earth or, so far as possible, built from aluminum and other metals taken from the Moon. There would be no cars (everything would be within walking distance), and sophisticated air-conditioning equipment would remove pollution while keeping temperatures comfortable. The sky would always be blue. Nor would the people lack for recreation, for the availability of zero-g would offer new possibilities. The colonists themselves would live in normal gravity, though, for their colony would rotate.

The colony itself, the Stanford torus, was conceived as a rotating wheel with six spokes and has a startling resemblance to von Braun's space station. It would be much larger, however: a mile across, providing a home for 10,000 people. Half the colony interior would be used for agriculture; the other half would serve as living space, divided into three communities, each with its own spoke. Living conditions then would be no more crowded than in some of the small European walled towns that date to the Middle Ages.

Again there would be a fleet of rockets to build it—the Flyback F-1. The first stage would be a winged version of the Saturn V first stage and would use the same engines. The second, cargo-carrying stage would be developed from elements of the space shuttle. Each such cargo rocket would carry two hundred tons to orbit at a shipping cost of twenty-five dollars per pound. Alternately, the payload compartment would be fitted out as a wide-body spaceliner, carrying two hundred passengers. For flight beyond low orbit, there would be a deep-space rocket, again developed from space shuttle hardware, capable of landing one thousand tons on the moon.

What would go to the Moon would be a lunar mining and transport facility. A crew of some one hundred moon-miners would be sent to scoop up lunar soil, package it in forty-pound bags, and then launch these bags into space using a mass-driver, or electromagnetic catapult. There in space, a large conical mass-catcher would serve to catch the bags, filling up at the rate of one hundred thousand tons per month. Once a month, a full load would be delivered to the colony.

Initially, of course, there would be no colony, but rather a "construction shack" equipped with processing and manufacturing facilities. Staffed with two thousand workers, and initially built in low Earth orbit, it would be moved to the colony site in deep space. There it would receive the loads of raw lunar material from the mass-catcher and process them into fabricated metal: aluminum, iron, magnesium, titanium. There would be oxygen, too, produced in huge quantities. Slag left over from the processing would serve to build a shield against cosmic radiation.

The function of the colonists would be to build solar power satellites: immense structures, miles in length and width, to convert sunlight to electricity and beam the electricity to Earth using focused beams of microwaves. At the ground these beams would be received and reconverted, each powersat providing enough electricity for New York City. In this manner, the existence of space colonies would help solve the energy crisis.

Such have been the dreams of those who would colonize space—imaginative and far-reaching, firmly grounded in present-day science and engineering, yet withal quite wistful and romantic. And how will these dreams look a quarter-century from now? Will the ten-thousand-person space community seem less naive, less unrealistic than von Braun's proposal to use two hundred astronauts to accomplish the military tasks that today are done entirely by automated equipment? Will the idea of space agriculture be seen as reflecting any real understanding of the actual problems in supporting a community of space workers?

We cannot answer these questions without an understanding of today's space programs and research and of the history of technology. The science of astronautics is older than most people think; it is arresting to realize that the first serious work in the field was contemporaneous with the Spanish-American War. It has seen recurrent waves of enthusiasm for various projects, the dreams of its visionaries. Time and again, its acolytes have come forth with enthusiastic speculations, offered confidently as predictions for the future. Time and again, these projects have failed to elicit the demonstrated practicality that would bring them to reality. Yet, astronautics has advanced.

The history of aviation provides a useful model for the future of astronautics. At the turn of the century, aviation was a risky and marginal activity of a few enthusiasts, having rather the same role in the nation as hang gliding today. (The technologies of the two activities are quite similar.) Today, tens of millions of people pass each year through such major airports as JFK and O'Hare, while such

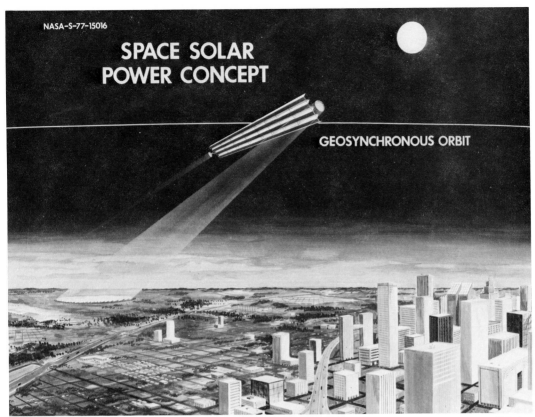

NASA-S-77-15016

SPACE SOLAR
POWER CONCEPT

GEOSYNCHRONOUS ORBIT

Concept of the power satellite. In geosynchronous orbit, 22,236 miles up, an immense space structure converts sunlight into electricity and, using large disk-shaped antennas, converts the electricity into a focused beam of microwaves which is transmitted to Earth. There a receiving antenna or rectenna reconverts the microwave energy to electricity. (Courtesy NASA)

once-impregnable institutions as the passenger railroad and steamship have long since been relegated to nostalgic supplements to a transportation industry dominated by aircraft. Thirty years ago, the departure of such liners as the *Queen Mary* or *Ile de France* was an Event; today, the Boeing 747 carries its hundreds of passengers at twenty times the speed, and its departure is noted chiefly by the air-traffic controllers.

None of this happened because of dramatic presidential decisions or of nationwide crash programs in response to foreign challenges. The key people who planned and carried out aviation's growth did not regard themselves as agents of destiny nor were they self-consciously driven by a sense of mission. The first flights of such revolutionary aircraft as the 707 (the ''Dash-80'') and DC-3 in no way were attended by the media coverage and ticker-tape parades that accompanied the first flights of astronauts in Project Mercury. Aviation's progress was the work of canny managers and corporate executives seeking business opportunity and the result of many small advances, few of which were in themselves revolutionary. Indeed, on the one occasion when commercial aviation indulged in a self-conscious sense of mission or destiny, the result was that magnificent economic flop—the SST.

55

Aviation is not the only activity offering instructive historical lessons to those who would colonize space. Over a century ago, the Age of Steam saw developments that appear today to be almost eerily parallel to the Apollo program. These events involved the building of the first large iron ships.

The sea has always been deeply conservative in its practice; innovations have traditionally been regarded with suspicion, as likely to fail and to cost the lives of sailors. The introduction of steam brought no sudden change in ships, either in their size or appearance. Well into the twentieth century, steamers continued to carry sail-bearing yards on their mainmasts. Far more so than in aviation, the development of modern navies and merchant marines was slow, gradual, deliberate, with change being all but imperceptible from year to year.

The advent of steam propulsion in the mid-nineteenth century stimulated the dreams of visionaries and attracted the talents of brilliant engineers, just as would the advent of rocket propulsion a century later. Foremost among these designers was one who may be regarded as the Wernher von Braun of his day—the Englishman Isambard Kingdom Brunel.

The son of Sir Marc Brunel, builder of the first tunnel beneath the Thames, Brunel distinguished himself early as chief engineer of Britain's Great Western Railway. When that railroad company expanded into shipbuilding in 1836, Brunel was chosen to design its steamship, the *Great Western*. It was of wood and was propelled by paddle wheels; at 1,320 gross tons, it was the largest steamer then afloat. It was the first ship to demonstrate the feasibility of operating a transatlantic passenger liner under steam.

His second ship, the *Great Britain* (1843), also broke new ground. Her capacity, 3,270 gross tons, made her the first large ship to be built of iron. Brunel invented the structural design used in her construction and built her so strongly that, even in 1970, her hull still could be salvaged from Sparrow Cove in the Falkland Islands and removed to permanent exhibition in Bristol, England. A commercial success, *Great Britain* convinced shipbuilding experts of the superiority of iron hulls for large ships.

These two ships came at roughly the same time in Brunel's career as did the V-2 and Jupiter-C rockets in von Braun's, and had the same impact on the shipbuilding world as did von Braun's projects in the world of rocketry. The V-2 of 1944 proved the feasibility of building large liquid-fueled rockets. The Jupiter-C of 1956 launched America's first satellite in 1958 and convinced critics of the desirability of using tested existing rockets whenever possible.

Von Braun's third pathbreaking project was Apollo. Brunel's was the *Great Eastern*. Apollo's principal rocket, the Saturn V, was an astonishing five times larger than the earlier Saturn I; and *Great Eastern* was five times larger than *Great Britain*. At 18,915 gross tons, with a length of 680 feet, *Great Eastern* represented a size that even today would be eminently respectable. In her day (she was launched in 1858) she was a leap decades ahead into the future.

Brunel built her expecting to exploit economies of scale. She could carry four thousand passengers and six thousand tons of cargo; she could run around the world, via Capes Horn and Good Hope, with only one refueling stop. He expected his ship would monopolize the sea route to India and Australia, but few shipowners shared his enthusiasm. During her construction, Brunel sought the advice of an expert friend: "If she belonged to you, in what trade would you place her?" The reply was direct: "Send her to Brighton, dig out a hole in the beach and bed her stern in it. . . . She would make a substantial *pier* . . . her old magnificent saltwater baths and her 'tween decks a grand hotel. . . . I do not know any other trade, at present, in which she would likely pay so well."

The Great Eastern, *largest of the iron ships of Isambard Brunel, foreshadowed the size and cargo capacity of transatlantic liners, but was built decades too soon to serve that trade. (Courtesy Garden City Publishing Co.)*

The Saturn V, largest of the rockets of Wernher von Braun, foreshadowed the size and cargo capacity of true space freighters, but was built decades too soon to serve that trade. (Courtesy Boeing Aerospace Co.)

Brunel ignored this advice and placed her in service. It soon was obvious that the cargoes he sought for her could not be found; the world of the 1860s had little demand for her immense capacity. She saw limited use on the North Atlantic run as a troopship to safeguard British interests in Canada during the U. S. Civil War. For a few years she was chartered to lay undersea cables, linking England and America by telegraph in 1866. By 1874 the advent of special cable-laying ships forced her out of that business, and she was laid up in port. Finally, in 1888 *Great Eastern* went to the breaker's yard. Not till 1899 did a larger ship, White Star Line's *Oceanic,* enter service. By then, advances in marine engineering had long since rendered *Great Eastern* obsolete.

It is impossible to resist noting the parallels to Apollo. The Saturn V and its Apollo spacecraft were built, not merely to go to the Moon, but to open up space to large-scale use. Yet the Saturn V failed to bring forth the cargoes or space traffic that could fill her huge capacity: 140 tons to orbit. By 1970 its production line was scheduled to be shut down, and the remaining craft were soon sent for display in museums. Attention shifted to the space shuttle, with advanced design features as well as a cargo capacity (32 tons) better suited to the traffic anticipated for the 1980s.

Yet while it took decades of time, that magnificent failure, the *Great Eastern*, indeed was succeeded by the successful *Oceanic*. And by the centennial of *Great Eastern,* 1958, the shipping world had long grown accustomed to far larger ships. Similarly, one may surely anticipate that in the decades to come, growing space traffic will compel anew the building of rockets with Apollo-size cargoes, and with larger cargoes still, but incorporating advanced designs, which will make the Saturn V entirely obsolete.

And if astronautics is to grow in this fashion—if a slow, patient advance will bit by bit increase the scope and tempo of space flight—then in time it will become entirely reasonable to undertake activities to challenge the imagination of space pioneers. Such activities will have their own pace and logic, and may no more resemble the classic dreams of space flight than today's cruise ships resemble Brunel's dream of a cheap, high-density passage to India. For all that, these space programs will be as startling, in comparison to the space world of 1979, as is JFK Airport in comparison to the aviation of the 1920s.

What may these programs be, and how may they come about?

In answering this, it is necessary to be cautious. The best sources of information come from the aerospace industry's studies and projections done by key people under contract. But such sources must be weighed and assessed with great care, for the aerospace industry is notorious for its self-serving tendencies. Charles Wilson, secretary of defense in the Eisenhower administration, once gained fame for allegedly remarking that "what's good for General Motors is good for the country"*; in the aerospace industry, this spirit is alive and well.

Thus, aerospace spokesmen have frequently justified space expenditures on the ground that they lead to spin-offs. A spin-off is a product or process useful in the nonaerospace economy, which was first developed or invented to support the space program. Such a spin-off is the Pillsbury food stick (originally marketed as Space Food Sticks), first developed by Pillsbury under NASA contract as a food for astronauts. A more substantive example of a legitimate spin-off is the Boeing 747. In 1965 Boeing and Lockheed competed to win an Air Force contract to build a large military transport aircraft, the C-5. Lockheed won the contract, but Boeing, not to be outdone, modified its design for the civilian market and put it into production as the 747.

---

*The correct quote: "I have always said that what's good for the country is good for General Motors, and vice versa."

Boeing's jet nevertheless was a spin-off from Air Force work. An often-cited spin-off from NASA research is the familiar Teflon-lined frying pan. Teflon, it has been said over and over again, resulted from NASA work; ergo, NASA deserves more money, the better to create more such wonders. In fact, though, Teflon was first developed by the Air Force. It was introduced about 1955 to reduce friction in the engines of some of their propeller-driven aircraft. A few years later Teflon served to line the large molds in which were cast the charges of solid propellant used in the Minuteman missile. Neither application had any bearing on the space program.

To be fair, there are legitimate spin-offs from NASA work that have use in such areas as medicine and fireproofing. But it could hardly be otherwise. It is difficult to see how the nation could spend billions on space projects without producing at least some spin-offs. In the main, though, space system requirements have been too uniquely specialized to produce inventions having any obvious use outside the space program.

During the heyday of the Apollo program, its contractors regularly invoked more than mere spin-offs as justification for the project. Their spokesmen often claimed that the lunar landing would produce a major scientific benefit: an understanding of how the world began. They expected to find that the Moon was composed of primordial, unaltered material, dating to the earliest days of the Solar System. In fact, however, it was early found that the material of the Moon had been melted, separated, and drastically altered early in its history to nearly the same degree as the rocks of Earth. So far from learning the origin of the Solar System, the Apollo flights failed to give even a firm indication of the origin of the Moon.

The industry response to this contretemps, predictably, was to call for more Apollo flights. The search was on for a "genesis rock," a truly primordial artifact, which might lie just beyond the next crater. This went on till the last Apollo flight, *Apollo 17,* in 1972. By then, Apollo-funded experimenters were turning in outstanding scientific work, but each Apollo flight was costing $400 million. The entire 1973 budget for the National Science Foundation, which supports most U. S. basic science, was $480 million. Measured against the priorities of the whole of basic research, the search for lunar genesis rocks simply could not compete. In any case, nonlunar genesis rocks could be found already, in museums. Certain meteorites, known as carbonaceous chondrites, indeed are truly primordial. The Allende meteorite, mentioned earlier, is an example.

The first Apollo landing was in 1969. Paradoxically, that year in fact did see a major advance in our understanding of the origin of the Solar System; but it had nothing to do with Apollo. It came in the form of a disarmingly slim book of mathematics, written in the Soviet Union by V. S. Safranov and published in the West under the title, *Evolution of the Protoplanetary Cloud and the Origin of the Earth and Planets.* It described the work of the Soviet school of planetary scientists, whose work in many respects surpassed that of Western investigators. Since then, that book has served as the point of departure for much of the best work on planet formation; its methods and approaches are regarded as entirely essential in modern studies of that problem. But such studies do not require flights to the Moon. They rarely require flights to the next state, unless a scientific conference is being held there.

It is essential to keep a skeptical attitude toward the claims of the aerospace industry. If indeed a space project is to go forward, the expertise of the industry is essential, but in light of the number of projects possible the key question is whether a particular project indeed should advance. It is in this skeptical spirit that one must also approach the statements that space colonization is inevitable, or that major increases in space activity lie on the horizon. Yet to those who advocate space colonization, such skepticism can be not a bad thing but an advantage. A skeptical approach will give full weight to difficulties, will recognize alternative approaches, and will emphasize an understanding of how large projects really come into being. If space colonization fails these tests, if it falls apart at the first serious attempt at criticism, then it is better discovered now than later. But if, despite such challenges, space colonization truly is found to be plausible and reasonable, then one can have high confidence that in time it will come about.

It is in this spirit that one may search for a reason to spark a vastly larger space program. A popular choice has been space industrialization: the building of factories in space to produce important new products by taking advantage of zero-g. Some advocates have claimed that this would usher in a "third industrial revolution," producing major changes across the entire range of modern technology. A study by Science Applications, Inc. has suggested that such activities could spawn a $6 billion per year manufacturing industry within the next quarter-century. Milan Bier, a leading specialist in biophysics, has gone further and suggested that a potential market of this size exists for just one type of space manufacturing alone—the purification of pharmaceuticals used in medicine.

Such pharmaceuticals would include a variety of extremely powerful hormones recently discovered through research on the brain. The most important of these are super-hormones, which do not directly control body processes but which regulate the production of other hormones that do. These super-hormones include LHRF (luteinizing hormone releasing factor), which regulates production of fertility hormones. There is also somatostasin, which regulates the action of the pituitary gland, controller of human growth. It is these brain hormones, which transmit the instructions or commands that shape the features of a life: "Keep warm," "Reproduce," "Grow no more."

Other important brain hormones are the enkephalins and endorphins. These are, to use Karl Marx's phrase, the opiates of the masses: morphine-like substances naturally found in the brain. Enkephalins serve in transmitting impulses along nerve pathways. The function of endorphins is not well understood, but it is believed that they aid in influencing the functioning of the pituitary. The study of these hormones has already led to major advances in fundamental understanding of the way pain-killers work, as well as the cause of drug addiction. Further work in this area is expected to bring not only better pain-killers, but also a cure for drug addiction.

Such hormones are very powerful; doses in the millionths of a gram would be typical. Yet their production would also be quite costly, in the tens of thousands of dollars per gram. Thus, if space processing would ease their production, the cost of the necessary space factories could readily be accepted. Such factories would rely on the production process known as electrophoresis, and it has been frequently claimed that the zero-g conditions of space would greatly enhance the usefulness of this process.

Electrophoresis is a powerful means for separating out the different components of a mixture of biochemicals. In this technique, electric charges are placed on the molecules of the mixture within a container filled with fluid. An electric field is set up across the fluid, causing the individual molecules to migrate across. Differences in the sizes or shapes of the molecules lead to differences in their rates of migration, so that a desired type of molecule can be separated out and purified even if it is present in minute concentrations. But the electrophoretic process is quite sensitive and easily upset by convection currents in the fluid. In zero-g such convection currents are absent. Hence some investigators have predicted that the size of electrophoretic apparatus could be greatly enlarged in space, leading to production rates thirty to one hundred times greater than on Earth.

Another possible product for a space factory is the large, highly uniform crystals of semiconductors, which are the basis for the modern electronics industry. It has been claimed that the absence of gravity would allow the growing of such crystals with greater purity and uniformity, greater freedom from imperfections, and larger size. Such crystals, in turn, might permit manufacture of improved solid-state devices. To support these ideas, advocates of space processing have made use of the results of crystal-growing experiments conducted aboard Skylab. In these experiments, space-grown crystals indeed were superior to crystals grown in similar experiments on Earth.

The prospects for space manufacturing were reviewed in 1978 by a committee of the National Academy of Sciences, the nation's foremost scientific body. The director of the review was William P. Slichter, executive director for materials science at Bell Laboratories. The general tone of their report was not at all favorable to the idea of space factories. To the contrary, their report again and again suggested the well-tried patience of specialist experts who had heard too many enthusiastic claims from people who didn't know what they were talking about. On the general idea of a program to develop space manufacturing operations, they wrote in "Materials Processing in Space":

> The early NASA program for processing materials in space has suffered from some poorly conceived and designed experiments, often done in crude apparatus, from which weak conclusions were drawn and, in some cases, over-publicized. Nevertheless, there is opportunity for meaningful science and technology developed from experiments in space *provided that problems proposed for investigation in space have from the outset a sound base in terrestrial science or technology and that the proposed experiments address scientific or technical problems and are not motivated primarily to take advantage of flight opportunities or capabilities of space facilities* [italics theirs]. . . . The identification of programs for investigation must be made by peer review, not by the availability of funds or the need to use a space facility.

On electrophoresis in space:

> The results of earlier experiments in electrophoresis in space are tenuous. . . . The objective of learning more about how electrophoresis apparatus should be designed and how gravity may affect the electrophoretic process will best be answered through well-planned terrestrial research rather than experiments in a low-gravity environment.

On improved materials for electronics:

> It is impossible to extrapolate the results of the [Skylab crystal-growing] experiments to specific materials or processes used or planned for use commercially or to predict any specific advantages of processing those materials in a low-g environment . . . It has been said that better starting material leads to better [electronic] device performance, but . . . most fabrication processes for devices . . . introduce physical and chemical defects far in excess of those originally present.

On the whole idea of space factories, their conclusion was unequivocal:

> When gravity has an adverse effect on a process, stratagems for dealing with it can usually be found on earth that are much easier and less expensive than recourse to space flight. . . . The Committee has not discovered any examples of economically justifiable processes for *producing* materials in space and recommends that this area of materials technology not be emphasized in NASA's program.

62   Why should this be? The people who advocate space processing or space manufacturing are not wild-eyed incompetents nor are they fools; to the contrary, they are sober and serious experts in their fields. The problem is that their field usually is aerospace engineering, which is not quite the same thing as pharmacology or materials science or solid-state electronics. Therefore, the scope of the aerospace engineer is limited to possibilities in space, whereas the pharmacologist or electronics expert may be attuned to alternatives that may be more effective than the space-oriented ones.

In light of this fact, one may look with better hope to those areas in which there already is a long and close partnership between aerospace and its customers. Then the customers come to the engineers to seek aid in new projects. The most outstanding example of such a partnership is the communications satellite. Can space communications lead to a very large space program? The answer is yes. Moreover, it will be nothing new if space communications pace the growth of the space program in the 1980s. It will instead be a continuation of past trends.

There can be no doubt that for all its postponed dreams and budget cuts, the space program has grown in twenty years from a highly dramatic sport for superpowers to an unheralded but highly significant feature of the modern U.S.A. This has happened in part—but only in part—through the Apollo program, one of the few major proposals of the space dreamers to be rendered into hardware. For the most part, the advances have involved small, undramatic steps in observations of weather and earth resources, military reconnaissance, astronomy and other sciences—and communications. As with the growth of aviation or the computer industry, the cumulative sum of many small advances has turned out to be a new and pervasive technology, largely taken for granted, yet influential in the lives of nearly everyone.

While many of the space dreamers wrote of flights to Mars or of space stations, the real advances of astronautics were going forward quietly, foreseen by few. Thus, in his 1957 book *Rockets, Missiles, and Space Travel,* Willy Ley had mentioned the possibility of placing satellites in what is now known as geosynchronous orbit, 22,300 miles up:

> [Geosynchronous orbit] holds a certain fascination. An artificial satellite in this orbit would need precisely one day for a complete revolution . . . it would appear motionless, always occupying the same spot in the sky like a fixed star. But aside from this fact the 24-hour orbit would not be very practical . . . The farther the orbit is from the earth's surface, the more fuel is required to reach it from the ground; from the point of view of fuel expenditure a lower, nearer orbit is more advantageous. . . . All factors are in favor of a low orbit.

Yet by 1972, fifteen years later, space traffic projections showed 43 percent of all traffic in the 1980s would be heading for geosynch. The advantage of having communications satellites fixed in position was so strong that no alternate orbit would do. Was the orbit difficult to reach? The aerospace industry responded by developing advanced rocket craft, particularly by building improved versions of the well-proved Delta launch vehicle. Today, looking to the 1990s and beyond, the communications industry faces a problem of which Willy Ley never dreamed: an actual *saturation* of the geosynch orbit, or shortage of desirable satellite locations along its arc.

Such saturation does not mean—at least not yet—that geosynchronous orbit will be as packed with spacecraft as a freeway during rush hour. Instead, it means that the satellites would be so closely spaced as to interfere with each others' operation. As of 1978 there were already some ninety-three spacecraft operating in geosynchronous orbit, or planned for launch. This rather lengthy list merely included the satellites that were planned to be in day-to-day use. It did not include such famous old satellites as Early Bird and Syncom, which pioneered the use of that orbit but which have long since been shut down as obsolete.

Satellites in geosynch are not evenly distributed, but tend to be clustered where they can view areas of heavy communications traffic. Locations over the western U.S., above the Atlantic, or over central Asia are particularly favored. If two such satellites operate at different radio frequencies, they will not interfere with each other no matter how close their locations. But there are only a few standard frequency bands used in communications. If two satellites are using the same frequency band and are closer together than 2° in longitude along the orbit, then there will be interference. Why? Because the radio beam sent up from a ground station spreads out as it rises, so that the ground station would be transmitting to two or more satellites instead of just to one.

So the world's growing needs for satellite communications cannot long be met simply by launching more satellites of existing types. Nor is there doubt that these needs will continue to grow rapidly in the years ahead. In 1978 there were 470 million telephones in the world, of which 165 million were in the U.S.; by the year 2000, the world total may top 2 billion. That year will also see over a billion TV sets in use throughout the world. In this rapid increase in communications traffic, satellite communications will set the pace. The amount of traffic carried by Intelsat, the international satellite communications organization, is expected to double every three and a half years.

The result of this challenge will be the obsolescence of the communications satellite as we have known it: a single compact spacecraft, launched aboard a single rocket and unfolded or deployed in orbit. Instead, there will be the communications platform. The first such platform may be under way as a formal project as early as 1981 and operating by 1986.

Its sheer size will put it in a class by itself. The largest communications satellite orbited to date, ATS-6, featured an antenna 30 feet wide. By contrast, the geosynchronous platform is proposed to be 269 feet by 102 feet in dimensions. It will do the work of many smaller satellites, thus relieving the problem of orbit crowding.

Each such platform will require five flights of the space shuttle. The first two flights will each carry to orbit an external tank, the large propellant tank that carries the shuttle's fuel supply. These tanks will be equipped with small rails and will be joined end to end while in orbit, forming a strong scaffolding on which construction takes place. One of these flights also will carry a twenty-five-kilowatt solar power system, developing more power than the whole array of solar panels used on Skylab. This power system will provide energy for construction of the platform and then serve to operate the platform once it is complete.

The third shuttle flight will carry a crane, to be fitted to the rails of the two-tank scaffold. It will also carry the structural components for the largest single communications antenna, a hundred feet in diameter. During this seven-day flight, astronauts will assemble the antenna and mount it to the scaffold with the aid of the crane.

The fourth shuttle flight will build the main, 18,000-pound platform structure. To do this, it will carry rolls of sheet aluminum as well as standard lengths of aluminum sections resembling angle irons. The shuttle also will carry a beam-builder: an automatic machine that cuts and welds the aluminum to form structural beams, lightweight and strong. The mission should last thirty days, time enough for the crew to form the beams and assemble them using the crane. The crew also will install prepackaged electronics and other, smaller antennas, thus completing the platform proper.

The fifth and last shuttle flight will carry a rocket stage to boost the completed platform to geosynchronous orbit. The boost must be prolonged and very gentle to avoid straining the large,

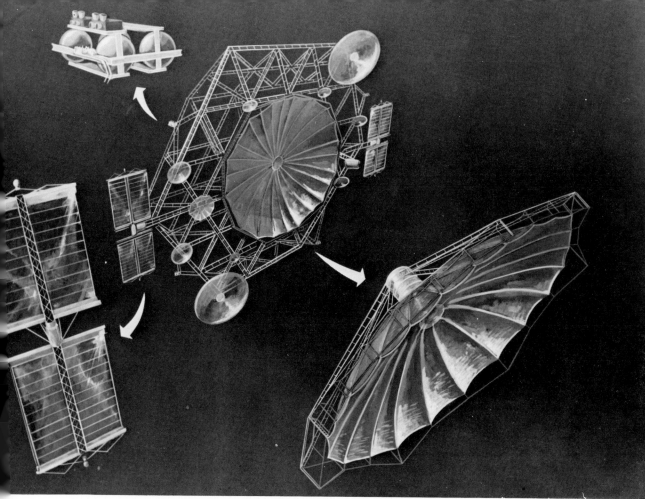

Early space platforms will serve the role of communications satellites. This one, studied at Marshall Space Flight Center, is 269 feet long by 102 feet wide. Solar panels, main antenna, and thrusters are shown in detail. (Courtesy NASA)

delicate platform with too-rapid motion. Thus, this transfer rocket will be a new design and a fairly major project in itself. This rocket will serve as well in future years, for as communications needs grow, the platform will also grow. Additional shuttle flights, supported by the transfer rocket, will deliver new electronics packages to the geosynchronous platform. There, automated robots—another new system—will install them.

In addition to providing expanded service for present-day communications, such platforms will offer entirely new services. Ivan Bekey, of the Aerospace Corporation (and now of NASA), has pointed out that as communications satellites grow very large, they can also be made very powerful, and able to service millions of ground stations. These stations, in turn, can be quite modest in transmitting power; yet their signal still will be picked up in orbit and rebroadcast. Carried to their logical conclusions, these ideas can give realization to an old fantasy: the Dick Tracy-style wrist radio, or telephone.

The wrist telephone would resemble a modern wristwatch; already some digital wristwatches

65

Use of the space shuttle to support early work in space construction. Forward of the shuttle orbiter is an external tank fitted with a solar panel and a crane running on tracks. The device with Mickey Mouse ears is a beam-builder, which fabricates structural members from rolls of metal. (Courtesy Grumman Aerospace Corp.)

incorporate the functions of a calculator, and this new service would be easy to add. There would also be a speaker, a small internal antenna, and a battery. Each wrist unit would have its own phone number and would respond with a ''beep'' when that number was called by signal from the satellite.

Such wrist units would not replace today's telephones; their tinny-sounding speech and use of a battery would make them suited only for short messages. Yet how often this would prove convenient! How commonly do we go on a trip, or one of our friends goes, and there is no straightforward way to keep in touch. Anyone who wanted to reach a friend would merely signal to his phone number, and if the number was right and he were wearing his wrist unit, he would be located anywhere in the world.

The wrist phones would be most useful for paging: ''I miss you; please get to a regular phone, and let's talk.'' In addition they would be very valuable for people who face danger. A hang-glider pilot might be blown to a remote canyon, but with his wrist phone he could guide rescuers. People who travel a lot, including long-haul truckers, might regard them as better than citizens-band radio. Such services then would offer still more opportunities for large communications platforms.

Such platforms will still fall far short of space colonization, yet will be far more advanced than previous space missions. Moreover, this project will prove out and develop many key techniques that may ultimately serve in the building of true colonies. For the first time, multiple shuttle flights will be needed to assemble a structure in space. For the first time, astronaut work crews will assemble these structures. The beam-builders, the use of a scaffold-mounted crane, the automated robots, and the advanced transfer rocket will be new. Most important of all, these new methods and techniques will not come about through speculation, or merely because key people are very taken with these ideas. They will come about entirely because of growing needs for communication. Their development thus is as inevitable as next month's phone bill.

Are there other activities, other than communications, that can provide the foundation on which to build an even larger space program? There are, and one of these involves a somewhat unpopular but inescapable matter, the military in space. Fortunately, what is in prospect here is nothing so gross (and, by treaty, illegal) as orbiting nuclear bombs or stationing missiles in space. What is to come instead is the next revolution in weaponry, one which may make the ICBM obsolete. As with communications, a bit of history again is in order. Even before the first A-bomb, writers of science fiction could look ahead to a defense against it. This was the death-ray, a powerful and concentrated beam of energy, easily capable of destroying a bomber or missile. To anyone steeped in Buck Rogers, a prediction in the 1930s of a mass attack by nuclear ICBMs might well have brought the riposte: Why

A large multibeam communications satellite such as might serve Ivan Bekey's concept of the wrist telephone. (Courtesy Rockwell International Corp.)

not protect against this science-fiction attack with an equally fictional death-ray defense? No attacking missile would serve any purpose if it could be blown up while in flight, before reaching its target.

The irony is that in the 1930s, such a viewpoint would have been entirely understandable. Years before there was any understanding of how to build a nuclear bomb, physicists in laboratories were already producing beams of protons and of electrons for their experiments. If not quite death-rays these beams nonetheless required careful handling since they were dangerous to life.

In the 1940s, the first atom bombs changed everything. With the development of powerful rockets and guidance systems, the threat of a nuclear-missile exchange emerged from science fiction to become a dominant reality in the world of the fifties and sixties. So destructive were these weapons, so terrifying the disaster that would follow their use, that nothing more than a change in their numbers or patterns of deployment could trigger a major international confrontation. That was what brought the Cuban missile crisis in 1962.

After that crisis, the world settled into a nuclear balance of terror. Neither superpower could restrain the development of improved weapons; neither could achieve any significant advantage over the other. Like a dormant earthquake fault, the nuclear threat was always there—not something to worry over in day-to-day life, yet not something to be dismissed and put out of mind.

Throughout all this, time and again there were people who suggested that the physicists' particle beams could be turned into weapons. The invention of the laser, a source of concentrated light beams, stimulated new thoughts of shooting down bombers or missiles with what would then be a life-ray. Yet none of these weapons proposals were convincing. The energy sources that would power them were inadequate; the beams could not be made powerful. Also there were serious problems of aiming, of detecting targets, and of causing the beam to propagate without spreading out or dissipating. In the early seventies the Navy sought to develop an electron-beam weapon (Project Seesaw). The abandonment of this effort, due to technical difficulties, only strengthened Pentagon reluctance to take seriously the idea of energy-beam weapons.

In the meantime General George Keegan, head of Air Force intelligence, argued forcefully that the Soviets were developing proton-beam weapons as a means of missile defense. The resulting debate was secret, but the implications for American defense potentially were quite strong. Keegan not only argued that intelligence data supported his views; he used Air Force funds to sponsor the research of a group of scientists and intelligence analysts at Wright-Patterson Air Force Base in Ohio. He emphasized his view that this group's findings clearly pointed to the feasibility of beam weapons. However, to the CIA and to the Air Force's scientific advisors, Keegan's views were unconvincing.

The side that lost out in the secret debate was the side that went public. Keegan leaked his intelligence findings to *Aviation Week,* a widely respected aerospace magazine. In its issue of May 2, 1977, editors Robert Hotz and Clarence Robinson warned that the Soviets had made a key breakthrough and were years ahead of the U.S. in particle-beams weapons research. The administration could not ignore such statements. The CIA issued one of its infrequent official announcements to deny that the nation was in danger. President Carter himself was moved to comment: "We do not see any likelihood at all, based on our constant monitoring of the Soviet Union as best we can, that they have any prospective breakthrough in the new weapons systems that would endanger the security of our country."

Yet in the fall of 1978 in a six-part series of articles in *Aviation Week,* plans were revealed to set up an Office of Directed Energy Technology in the Defense Department. The Navy would have principal responsibility for electron beams, the Army and Air Force jointly for proton beams, and the

Army for beams of neutral or uncharged atoms. The Air Force also would carry forward with development of high-power laser weapons, emitting intense, highly focused beams of light. One could imagine a hydrogen atom marked ARMY with its nucleus (a proton) marked AIR FORCE and its outer electron bearing the legend U.S. NAVY.

The Navy's program was revealed as the most advanced and having the highest priority. Called Chair Heritage, it currently looks toward key experiments in late 1981 or early 1982 to determine whether electron-beam weapons can protect cruisers and aircraft carriers from Soviet cruise missiles. Chair Heritage is being developed first because naval vessels can already provide energy from on-board power supplies to operate particle weapons. Their propulsion turbines develop high levels of power. Also it is easier to defend against cruise missiles than against an ICBM.

The Navy concepts call for individual electron-beam devices to be installed below decks. The beams would be guided by magnets and fired from rotating cylinders. These pulsed beams would make use of key findings in the Chair Heritage program: An electron pulse bores a hole in the air, leaving a trail which is hotter and less dense. Electrons in the next pulse face less resistance as they travel toward the target. However, this hole does not act as a channel. The beam can be moved and will carve a new path to the target. Also, it has a "self-pinching" effect, which keeps it from spreading out.

The Air Force is seeking to develop powerful lasers to protect its reconnaissance spacecraft from Soviet killer satellites. In the past decade the Soviets have conducted at least sixteen antisatellite tests. During ten of them the killer satellite passed within a few thousand feet of its intended target, close enough to ensure a successful kill. An antisatellite laser would fire at the Soviet spacecraft from the ground, blinding its sensors and antennas. With sufficient power, the laser would vaporize part of the satellite structure, causing further damage.

Can such lasers or particle beams be placed aboard orbiting spacecraft? An obvious possibility is the "Trojan powersat": a power satellite that in time of peace would provide electricity to the ground, but that all along would secretly carry beam weapons as armament. Like the Trojan horse of Homer's *Iliad*, in time of war it would prove a gift of which to beware, for it would easily supply enough power to provide ample defense against an attack on its home nation.

The problems with this are several. Powersats will be so large and flimsy that they will not readily be protected against attack; they could easily be disabled by other beam weapons. In war such armed powersats would prove attractive targets, for whatever their value as producers of energy, they would first have to be destroyed before an attacker could launch his missiles with confidence. More than one nation might build such armed powersats. In time of international crisis, a battle of dueling powersats might knock out much of the world's energy supply.

An armed powersat would actually be a latter-day version of one of the weapons of World War I. At the turn of the century, the pride of Britain and Germany were their fleets of luxurious transatlantic liners. These were advertised as "greyhounds of the sea," but what was not advertised was that they were equipped to carry naval cannon in wartime. The *Lusitania,* for one, could mount twelve six-inch guns, thus delivering a heavier broadside than the cruisers that guarded the English Channel. When a German submarine sent that ship to the bottom in 1915, the English press fell over itself in denouncing this as a horrible act of barbarism. What was not publicized was that in the 1914 edition of Britain's standard naval reference *Jane's Fighting Ships* (issued to all German sub commanders), the *Lusitania* was listed as an armed auxiliary cruiser.

This event demonstrated the futility of arming important civilian ships as vessels of war, and it has not been done since. One hopes that future generals will remember this history and not cast covetous eyes at the powersats. As with the world's navies, specialized spacecraft will be needed.

For defense against ballistic missiles, the Army is pursuing its space-based Sipapu program (an American Indian word meaning "sacred fire"). Sipapu calls for orbiting generators to produce intense beams of neutral hydrogen atoms, which would not spread out in space as would beams of protons or electrons. The Army is also interested in orbiting lasers. Such lasers could deliver concentrated pulses of energy a million times more intense than the concentration of energy from a nuclear bomb. What's more, the laser can deliver its energy in one-millionth the time of a nuclear weapon. Yet the laser is not a weapon of mass destruction, capable of annihilating cities and ports. It is a weapon of discrete destruction. It offers extremely precise control of rather small amounts of total energy, such as the energy of fifty pounds of high explosive, sufficient to destroy an enemy missile. If an ICBM is killed just after launch, it will stop accelerating. Its warhead may land in the home territory of its own nation.

Quoting officials close to the Army's Sipapu and laser programs, *Aviation Week* (October 16, 1978) had this to report:

> When lasers are placed in space so that every location on this planet is placed continuously in the target area of a laser battle system, then one has the right to expect truly fundamental changes. It raises the distinct possibility that the rapid delivery of nuclear explosives can be prevented by a weapon system that is itself not capable of mass destruction.
>
> When space transportation attains sizeable economies, then space weaponry must be evaluated on the basis of military utility rather than being summarily dismissed because of huge logistics costs.

Scenes of space war. Top right, an observation satellite has detected the launch of enemy ICBMs. Top left, other satellites track the missiles while in flight. Foreground and right, orbiting particle-beam weapons of the Sipapu type produce intense beams of neutral hydrogen atoms, which blast the missiles out of the sky. (Painting by Larry J. Herb, courtesy Aviation Week & Space Technology. © 1978 McGraw-Hill Inc.)

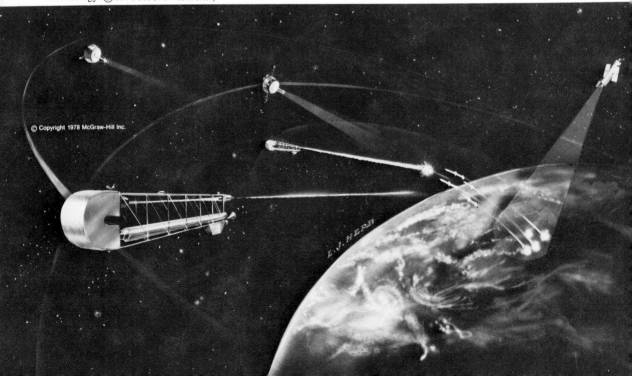

Such weaponry need not be placed in primitive, flimsy satellites. Rather, heavy weights of shielding and hardening materials become feasible in space. The term battle station is more descriptive of these weapons than the images conjured up by the terms satellite or space station.

A few dozen satellites are easy to envision, but a few hundred also should not shock us. The average naval vessel is much larger and we consider that 1000 of them are necessary to our peace of mind. We could doom ourselves forever simply by not realizing the fabulous results of applying some ingenuity to space logistics.

If space battle stations are to be built, hardened against attack, they may come to be built from lunar resources. Lunar soil would provide a bulky material for thick, attack-resistant walls. There may be spacecraft resembling the battle star of *Star Wars,* with immense shells full of lunar materials.

Beyond this, it is unwise to speculate. We are as if we had lived a century ago, trying to imagine the future of World War I. In 1880 war was still regarded as chivalrous and grand, seen in terms of Napoleon and *la gloire*. Only as recently as 1862 Robert E. Lee had watched the lifting fog at Fredericksburg disclose the massed Union host and had exulted: "It is good that war is so terrible, else we should grow too fond of it." As we look ahead, our parochialism is similar.

The matter of military activities in space is shrouded in secrecy and thus is difficult to assess. The same is not true of what may offer the most important prospects for a really extensive space program. These prospects lie in the area of energy. Indeed, it is possible that we will seriously look to space for a long-term solution to our energy needs. We then would reach into space so extensively that amid our activities, space colonization could come about in an entirely natural way.

The idea of the power satellite began as a speculation, and had it remained so, space colonies today would appear as one speculation founded upon another. However, in recent years the world has changed. Even very speculative energy sources have received new and often serious attention. It thus is appropriate to recall how these changes in the energy picture have come about, and not only give a further description of the powersat, but also to assess its importance.

CHAPTER 4

Energy, at Leisure

Early in the morning of October 6, 1973, Egyptian troops stormed across the Suez Canal and cracked the Israeli defenses. Eleven days after the outbreak of war, representatives of the Arab oil-producing nations met in Kuwait. There they agreed to cut back production of oil and to embargo its shipment to the U.S., as well as to the nations of Europe which were friendly to Israel. The embargo was never total, and in March 1974 it was lifted entirely. But within those few months, the world had changed. The result of that epochal meeting in Kuwait was not only the energy crisis, which saw motorists getting in line at five a.m. at gas stations which would not open till eight. There was the looming realization that cheap and accessible energy could never more be taken for granted.

The end of the cheap-energy era came quite suddenly. In 1947 the Persian Gulf's oil reserves had come under the control of an Anglo-American consortium of oil firms, which set the price of crude at $2.17 a barrel. The price stayed at that level till 1959, when the oil cartel cut it to $1.79. This price cut led the Arab nations to organize a counter-cartel in 1960, which they called the Organization of Petroleum Exporting Countries (OPEC). For thirteen years, despite increasing control over the oil resources in their lands, the OPEC ministers were unable to budge the price of oil from that level of $1.79; but at the Kuwait meeting they made their move. The price went up to $5.12 a barrel.

In December 1973 the Shah of Iran, an advocate of much greater price hikes, held an auction of oil on the spot market, selling individual tanker-loads of oil not committed to long-term contract. When some sales went as high as $17 per barrel, the Shah knew he could safely push for what would be a large hike indeed. By early 1974 he had led OPEC into a price rise to $10.95 per barrel, and there it stayed.

With the lifting of the embargo, gas lines disappeared, supplies once again became readily available, and prices steadied. These higher prices cut the growth of oil consumption, and in the mid-1970s Arabian crude was in oversupply. In these same years, important new oil fields were brought into production in Alaska and the North Sea, lessening world dependence on the OPEC powers.

However, this apparent return to normalcy was in fact merely an Indian summer of the old order, with a new winter of oil shortages only a few years off. The second oil crisis was triggered early in 1979, when Iran's Ayatollah Khomeini ordered a strike of Iranian oil production in order to build pressure for an overthrow of the Shah. When the Shah left the country and Khomeini was installed as head of a new government, he resumed oil exports, but at a level two million barrels a day less than before. This move, plus the interruption in exports during the strike, proved quite sufficient to take out all the slack in world oil production, putting massive, new, upward pressure on prices. Prior to Iran's upheavals, the OPEC producers had frequently sold petroleum at a discount from their posted price of $12.70 per barrel. After the crisis they raised the price to $18, but by restricting production they found few difficulties in demanding surcharges, which lifted the actual price per barrel to $22 or $23. The result: a return to the gas lines and station closings of 1974, with motorists paying over a dollar per gallon.

The roots of this second and likely more enduring shortage trace to 1956. In that year Egypt seized the British-controlled Suez Canal, touching off an invasion by the combined forces of Britain, France, and Israel. The U.S. secretary of state, John Foster Dulles, refused to support this high-handed action, and forced a withdrawal by the invading forces. This meant a defeat for the old, aggressive methods of imperialism and gunboat diplomacy; but in the long run it meant that the industrialized nations would become dependent upon the oil of small, weak, unstable nations, without having direct control in those countries. At the very moment when the old policies of imperialism and colonialism stood to reap their biggest dividends, these very policies stood discredited and impotent.

Also in 1956 the geologist M. King Hubbert first pointed out the limits on the U.S. oil supply. He asserted the U.S. would discover and produce no more than 150–200 billion barrels, and this meant domestic production would peak out by 1970 and thereafter decline. More optimistic estimates predicted a total of 590 billion barrels, which would push the date of peak production to close to 2000. Hubbert in turn replied that those estimates assumed we would continue to discover new oil at the same rate as in the past, but that in fact new discoveries had dropped off drastically. Indeed, by tracing this falloff, Hubbert once again was led to predict a total of some 170 billion, with a peak of production in the late 1960s.

Hubbert was right. Domestic production peaked out at 10 million barrels per day in 1970, and by 1973 was down to 9.3. In 1978, even with 1.2 million per day from Alaska, total production was down to 8.7 million and falling rapidly in the "lower 48" states. Daily consumption was up to some 17 million barrels, the difference being imported. In 1976 domestic oil reserves fell to 30.9 billion barrels; 2.8 were consumed. This supply was offset by 1.1 billion barrels of additions, mostly bookkeeping shifts from "probable reserves" to "proved reserves." Only 0.068 billion barrels were found in new fields. Nor were prospects bright for new discoveries in offshore fields. In the late seventies the major oil companies spent nearly $1 billion—the cost of their 1968 Alaskan leases—to gain the right to drill in a promising formation, the Destin Anticline, in the Gulf of Mexico. The result: dry holes. Much the same happened in another formation, Baltimore Canyon.

The 1978 discovery of huge new reserves in Mexico sparked hope that that nation might become a major oil producer. In the long run, it probably will, but that oil is under the control of the domestic oil monopoly, Pemex. Among oilmen, Pemex has a solid reputation for featherbedding labor practices, for waste and inefficiency; it is known as the only oil company ever to lose money. This may in part explain the policy of Mexico's president, Lopez Portillo, which is that that oil will be brought into production slowly, and exports kept to modest levels.

When President Carter came to office in 1977, he identified coal as the key fuel, the expanded production of which would alleviate our reliance in oil and natural gas. He called for America's production to double to 1.2 billion tons per year by 1985. However, this goal may be out of reach, and the problems with coal illustrate sharply the disarray in the nation's energy programs.

Since 1973 regulatory and legal delays have lengthened drastically. The time necessary to open a large coal mine has increased from five years to as much as ten or more. Many federal, state, and local agencies have set separate but overlapping requirements, which differ from place to place. As many as one hundred permits now are needed to open a mine, all of which must be obtained before production can begin.

A most valuable resource is the coal of the Rocky Mountain states. Much of Carter's proposed expansion in coal production is to take place there, through the development of enormous strip mines. In Montana, seven such sites are planned, with a projected output of seventy-five million tons per year. Western coal is lower in energy content than is coal from eastern mines and must be hauled longer distances to its users. But it has an important advantage: It is low in sulfur.

Sulfur is one of the worst pollutants in coal, and in 1971 permitted emissions of sulfur oxides were set by law at a very low level. To comply with the law, utilities had two choices. They could install complex and trouble-plagued pollution-control devices, known as stack-gas scrubbers, to capture the sulfur oxides after burning and prevent their discharge. Such scrubbers have increased the cost of power plants by up to one-third. Or, the utilities could burn coal that is naturally low in sulfur. Much western coal is low enough to comply with the law, so that its use might be expected to grow rapidly.

However, in 1977 Congress passed amendments to the Clean Air Act, which now require *all* new coal-burning facilities use stack-gas scrubbers. Thus, new coal-fired power plants cannot comply with the law by using low-sulfur coal; they must have costly scrubbers, too. This deprives western coal of its competitive advantage, which is the same as writing off prospects for its rapid development. Thus, if 1985 comes and U.S. coal production is lagging, it will not be because of an energy crisis. It will be due to a legal and regulatory crisis.

The situation is even less promising with nuclear power. The era of commercial nuclear electricity got under way during the Kennedy administration with the building of a number of plants of 200–300 megawatts capacity. These were similar in design to the well-tested reactors that powered naval submarines. By the late 1960s, with little hindrance from the Atomic Energy Commission, the nuclear industry was moving aggressively to promote and build a new generation of much larger plants. These plants, in the 800–1200 megawatt range, lacked the strong technical base of operating experience that had supported the earlier ones. To some nuclear critics, it was as if the airline industry had moved in one leap from propeller-driven aircraft to the wide-body jets, without an intervening decade of operating experience with jets like the 707 or DC-8.

Inevitably there were problems. The new plants were often out of service for repairs or maintenance. Increasing regulation of the nuclear industry meant that new plants could need up to twelve years from time of order to entering service; costs soared. By the mid-1970s utilities had lost much of their early interest in nuclear plants and had entered into a de facto moratorium by virtually ceasing their new orders. Still, the momentum of earlier orders meant that by early 1979 there were seventy-two plants in service, generating 13 percent of our electricity, with another ninety in various stages of construction. Chicago was getting half its power from the atom; New England, some 40 percent.

Then came Three Mile Island. On March 28, 1979, the nuclear plant near Harrisburg, Pennsylvania, suffered the worst reported accident to date in a commercial nuclear plant. This plant had been ordered in 1968, as one of the new generation, with a rated power of 865 megawatts. The accident started when a pump failure forced a ''turbine trip'' or generator shutdown. Auxiliary pumps were supposed to circulate cooling water to the reactor core, but could not do this because an operator had mistakenly left two key valves shut. In the control room another error took its toll: An important gauge malfunctioned, leading controllers to the mistaken conclusion that there was adequate cooling water in the core. Before things got back under control, the core had suffered massive overheating and partial melting, spilling very high levels of radioactivity into the containment structure, which housed the reactor proper. Some of this radioactivity leaked into the air, and many residents of the surrounding areas had to be evacuated. In the days that followed, it appeared to many observers that key people in charge, at both the Nuclear Regulatory Commission and the power-plant utility, simply had lacked adequate understanding of the systems they were trying to control.

This accident brought home to the nuclear skeptics the inherently risky nature of the enterprise and the need for better safety precautions. However, this accident did not mean the demise of the nuclear industry. Its long-term effect probably would be more nearly comparable to the *Apollo 204* fire in January 1967. As the spacecraft of that name was undergoing ground tests, a sudden fire broke out, taking the lives of its three astronauts. The result was a wholesale reorganization of the Apollo project, and a strong new emphasis on safety, particularly against fire hazards.

It is an inherently risky matter to generate electricity with nuclear energy. It is also inherently risky to put two hundred people in an airplane and send them hurtling across the sky at thirty thousand feet and six hundred miles per hour. In the aviation industries, safety is part and parcel of all design and operating practice, and pilots and crewmen have long experience and training. Today's challenge is for the nuclear industry to make a similar commitment to safety. To the degree it can do that, it will still have a future. Nevertheless, old dreams of a nuclear-electric America can hardly be sustained. Not only is there the safety question; there also is the availability of uranium. It is becoming more costly, and current projections suggest there will be only enough, at acceptable prices, to fuel two hundred large reactors for their thirty-year lives.

When President Carter was inaugurated, many of these events lay in the future, but the trend of our increasing dependence on Arab oil was plain. In April 1977 he addressed the nation on the subject of energy, calling on Americans to fight the ''moral equivalent of war.'' In this he was wrong; it was not the moral equivalent of war. It was, and is, the economic equivalent of war. But in facing these latest challenges, the U.S. is not without the means to cope.

It is the present policy of the Department of Energy to proceed with support of full-scale commercial synthetic-fuel plants, which by 1983 or 1984 will give us energy options we do not now possess. These options would take the form of proved and demonstrated technologies, which could then be expanded as needed.

Two of these options involve the conversion of coal into clean fuels by a process known as solvent refining. A third process will produce pipeline-quality natural gas from coal at a plant to be built in North Dakota. These initial plants will not in themselves solve the energy problem; the two solvent-refining plants, for example, will each produce the equivalent of only about twenty thousand barrels of oil per day, or a thousandth of our domestic requirements. Their importance lies in the new prospects they will offer. The U.S. has some 437 billion tons of coal that is currently mineable, enough to produce over a trillion barrels of liquid fuels. This is twice the total world reserves of petroleum, and six times the proved reserves of Arabia.

An even more significant synfuels industry may grow up around the oil shale of Colorado, Utah, and Wyoming. Those states contain vast deposits of shale, which is impregnated with a rubbery solid, kerogen. When heated by injecting air underground and lighting a fire, the kerogen turns to oil, which can be collected and pumped out. The total U.S. reserves of shale oil dwarf even the potential of oil from coal; these reserves amount to some two trillion barrels.

These synfuels industries will produce more than enough energy to tide us over to the era of permanent or renewable sources, but their products will not be cheap. Their cost will be some $30 per barrel at the refinery, and gasoline from these sources will cost $1.50 per gallon in today's dollars. However, the auto industry by then will be building diesel-powered cars, which will get fifty miles to the gallon. When fifty-mile-per-gallon cars burn $1.50 gasoline, it will be the same as in the good old days, when cars got twelve miles per gallon but gas was $.36 at the pump.

It is far too early to write off prospects for America. Time and again, Americans have gone ahead with business as usual, downgrading or ignoring challenges from overseas till they were galvanized into action by a sudden shock—a Pearl Harbor, a Sputnik. Then they have responded with vast and successful national efforts, confounding the critics who pronounced us too weak, too irresolute, too self-centered, or preoccupied with personal pleasures. It could well happen again and probably will.

And when, by whatever means, we finally begin to treat energy with the seriousness it deserves, we will go forward to develop permanent, renewable sources. These will last us not for mere decades, nor even for centuries, but for as far into the indefinite future as we will care to plan. When they are fully developed and operating, they will be the foundation for our civilization.

In recent years one such permanent source has dominated our energy research budgets: the fast-breeder reactor. In the breeder, uranium 238 is exposed to neutrons and is converted into an isotope of plutonium, Pu-239, which is fissionable and can produce power. The attractive feature is that the breeder not only produces power, it can produce more plutonium than it needs to keep running. It has been described as functioning like a soda machine in which you would put in a quarter and which would give you not only a soda, but thirty cents.

However, there is more to nuclear energy than just the production of power; and the breeder offers some of the worst side effects imaginable. Plutonium is separated out and purified by a chemical process known as Purex. The Purex process has a long and distinguished history; it was first invented to produce plutonium for nuclear bombs. So a fast-breeder reactor, with its facilities for producing plutonium, is nothing less than a factory for atomic weapons. Alvin Weinberg, former director of the nuclear laboratories at Oak Ridge, Tennessee, had this feature in mind when he wrote in *Science*, July 7, 1972:

> We nuclear people have made a Faustian bargain with society. On the one hand we offer—in the catalytic nuclear burner [breeder]—an inexhaustible source of energy. . . . But the price we demand of society for this magical energy source is both a vigilance and longevity of our social institutions that we are quite unaccustomed to.

Or course, to say that disasters could happen is not to say that they will. The U.S. military has for over thirty years safeguarded some one hundred tons of bomb-grade plutonium, much of it in the form of bombs. But Weinberg's "Faustian bargain" is quite real, and the most likely role for the breeder is as the energy source of last resort. As long as there are other prospects for achieving energy, the breeder will face most severe opposition and will be developed only with great reluctance.

Fortunately, the fast-breeder is not the only available permanent energy source. It is not generally appreciated that there is a way to build a nuclear plant whereby it will tap a virtually inexhaustible fuel supply, require no enrichment of uranium, operate quietly and routinely with almost no shutdowns, while producing at least half a million kilowatts of power. Such reactors indeed are working every day and have done so since 1967 in Ontario, Canada. These are the reactors of the CANDU (Canadian Deuterium Uranium) type.

CANDU reactors use their neutrons very effectively and economically and run on natural, unenriched uranium. The low natural fraction of fissionable U-235, 0.71 percent, is no handicap because these reactors use heavy water (deuterium oxide) as a moderator. All reactors require a moderator: a substance which slows down neutrons while absorbing as few as possible. U.S. reactors use ordinary water as a moderator, but heavy water is much better and absorbs far fewer neutrons. Hence CANDU reactors can operate with incredibly low fractions of U-235.

Present CANDU reactors, unfortunately, do produce plutonium. However, such reactors can operate using an entirely novel nuclear fuel, thorium, which is three times as abundant as uranium but which cannot be fashioned into bombs. A CANDU reactor, fueled with thorium, would use a little natural uranium as a source of neutrons; the neutrons would convert some of the thorium (Th-232) into another isotope of uranium, U-233, which is fissionable and which would produce the reactor's energy. Thereafter, the system would need no further uranium and only small amounts of abundant thorium; it would be nearly self-sufficient. To use the soda-machine analogy, it would not return a soda plus thirty cents, but it would deliver a soda plus at least twenty-four cents.

This is not the only hopeful prospect on the energy horizon. Within a very few years, the long-held dream of nuclear fusion should be well on its way to reality.

Serious fusion research started early in the 1950s, when such physicists as Lyman Spitzer used simple physical theories to propose that it should be possible to confine a plasma, a gas of charged particles, within magnetic fields, which would form a "magnetic bottle." If the plasma were a mixture of deuterium and tritium, the heavy isotopes of hydrogen, and if it were heated to sixty million degrees centigrade and kept there, then there would be produced energy reactions similar to those which power the Sun. A man-made star, a new source of power, would glow within the laboratory.

The first fusion experiments took place in such machines as Los Alamos' Perhapsitron (the name reflected the dubious nature of the enterprise) and Princeton's Stellarator (which expressed the hope of harnessing the energy of stars). The results were immediate. When the physicists turned on the power, the plasmas propelled themselves out of their magnetic bottles within a few microseconds.

Obviously, something was very wrong. By the early 1960s, it was clear that there were two major problems that stood in the way of success. The first was that plasmas in magnetic bottles were unstable. There was a depressingly long list of ways the plasmas could manifest instabilities; any one of these would prevent a magnetic bottle from confining the plasma.

The second problem was more subtle and involved the rate at which a plasma would leak from even a well-designed magnetic bottle. In the simple theory of the early fifties, the leakage was held to be governed by "classical diffusion," which would diminish rapidly with the increasing strength of the magnetic field. Hence modest increases in the field strength would aid greatly in controlling the leakage. But the fusion machines of the sixties showed "Bohm diffusion," first studied by the physicist David Bohm. Bohm diffusion diminished much more slowly with increasing magnetic strength, so that to achieve a properly low leakage rate would require an impractically strong magnetic field.*

These problems slowed fusion research to a crawl and spawned the legend that fusion would always be impractical barring a breakthrough. But the breakthroughs came. By the late sixties physicists armed with powerful new theories were conquering one instability after another. The true breakthrough came from the Soviet Union. In 1968 Lev Artsimovich, director of Moscow's leading fusion research center, announced test results from a new type of fusion machine, the tokamak (a Russian acronym for "toroidal magnetic chamber"). The tokamak did not show Bohm diffusion. This meant that if only a tokamak was built large enough, it would work. It would produce power.

Soon many physicists were building tokamaks or modifying existing machines. In November 1971 Robert Hirsch, director of the U.S. fusion program, went before a Congressional hearing and stated that with sufficient funding, the U.S. could have a working fusion power plant by 1995. The result of these new developments was that fusion funding, at less than $30 million per year through the

---

*To be precise, classical diffusion scales as $1/B^2$, where B is the magnetic field strength. Bohm diffusion scales as $1/B$.

Banks of lasers in the Shiva facility produce the intense bursts of energy which are focused upon pellets of deuterium and tritium in fusion experiments. (Courtesy Lawrence Livermore Laboratory)

Boeing's powersat design features an immense flat array of silicon solar cells. In the foreground is the transmitting antenna, whose rotary joint allows it to face Earth continually while the solar arrays face the Sun. (Courtesy Boeing Aerospace Co.)

sixties, suddenly took off. In 1974 it was $51 million; by 1978 it was $290 million, and heading higher.

This funding served to build new, large tokamaks for further experiments, and results were not long in coming. In 1977 MIT's Alcator machine set a record by attaining 50 percent of the plasma conditions required to produce net fusion power.* However, the Alcator results were at the comparatively low temperature of ten million degrees centigrade. In 1978 the Princeton Large Torus raised the plasma to sixty million degrees centigrade using a new method of heating, an important accomplishment showing that even at that high temperature the plasma was stable.

True fusion power, the condition of "breakeven" wherein a fusion experiment produces more energy than it needs to operate, will come with the next major experiment. This is Princeton's Tokamak Fusion Test Reactor (TFTR), a $230 million machine scheduled for completion early in 1982. So within a very few years, we can anticipate a dramatic announcement: the opening of the Fusion Age.

---

*The conditions are given by the Lawson criterion, which states that the product of plasma density (particles per cm³) and confinement time (seconds) must be 6×10^{13} or greater. The Alcator record was 3×10^{13}.

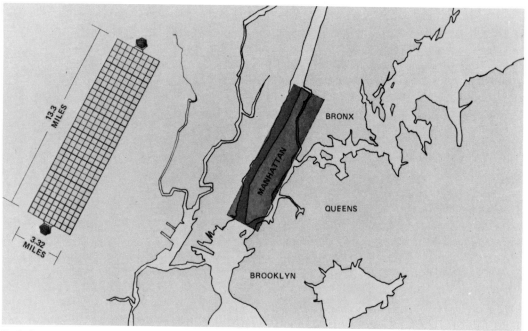

Boeing's powersat features a solar array 13.3 miles long by 3.32 miles wide, which is larger than the island of Manhattan. Powersat would deliver five million kilowatts from each of its two transmitting antennas. (Courtesy Boeing Aerospace Co.)

Nor will the TFTR be the last word in fusion reactors. Intensive work is already being done to devise advanced fusion machines, which will make more effective use of their magnetic fields and thus lower their costs. Other experiments are studying the use of fuels more advanced than deuterium-tritium. Additionally, there have been advances in an entirely different approach to fusion, in which a pellet of fuel is instantaneously compressed or imploded by being struck with powerful energy beams. The Shiva experiment, at Lawrence Livermore Laboratory near San Francisco, has already shown how a large laser can do this. Its successor, called Nova, will go further and reach breakeven by this entirely new route, a year or two after the TFTR. Other experiments, aiming at producing fusion using electron beams or beams of high-energy atomic nuclei, also are making important contributions. But the success of the TFTR will be the key event.

And with this achievement, the fusion program will reach a milestone that compares with the event which opened the Atomic Age: the first successful nuclear reactor in 1942. The Princeton experiment will not mean that fusion power will thereafter be lighting America. As with atomic power, it will mean that perhaps in fifteen more years—by century's end—the first fusion plants will be producing commercial power. In another fifteen years, say by the year 2015, fusion, like nuclear power today, will begin to make an important contribution to the nation's energy needs.

Success in fusion, then, may form the foundation for civilization in the next millennium; but it will not transform our energy needs overnight. This drawback raises the question of whether we will speed things up by relying on that big fusion reactor in the sky, the Sun. Solar energy has in recent years become quite popular, and there is no doubt it has an important role to play. It can heat homes, provide hot water, and serve a variety of uses where conventional electricity is too costly, as in remote areas. Solar energy can grow crops to be fermented to produce alcohol, which can be blended with gasoline. As its proponents never tire of noting, solar power is decentralized, democratic, available to all. Advocates such as Amory Lovins (of Friends of the Earth) have hailed solar power as the way to free the nation from dependence on the oil companies and other energy giants. They suggest the prospect of a world where everyone will have his own personal energy system.

There is nothing fundamentally absurd with decentralized, personal energy systems; most people have a decentralized, personal transportation system in the garage. But the history of decentralized energy is not encouraging. Many farmers and ranchers used to have windmills to generate power. But the windmills were costly and unreliable, and these people gladly welcomed the chance to hook into the nearest electric power network. Even today, few people propose to build a life-style around the ideal of self-sufficiency in energy, since it is much more convenient to pay the monthly electric bill than to wrestle with bulky generating equipment in the backyard. The joys of decentralized power, of Lovins' ''soft energy paths,'' somewhat resemble those of centralized transport, of many forms of mass transit. In both cases the advantages are more convincing to social planners or political reformers than they are to the average citizen.

There are a variety of projects for using the Sun to generate electricity on a large scale, but few of them carry much conviction. Furthest advanced is the ''power tower,'' which employs large fields of mirror reflectors to focus desert sunlight onto an elevated boiler. It may see limited use in some communities of the Southwest, but at up to fifteen times the cost of a coal-fired plant, it bids fair to be one of the most costly ways of producing electricity ever invented. Somewhat better prospects exist for large windmills, with blades as large as the wings of a 747; but these would largely be limited to windy areas in Wyoming and the Rockies. Most dubious of all is the proposed Ocean Thermal Energy

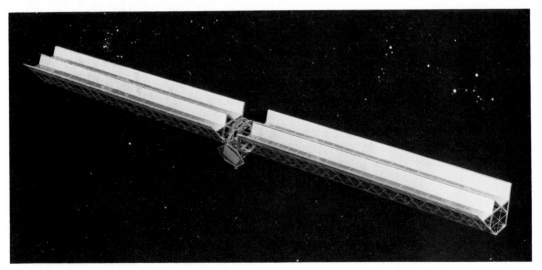

Rockwell International's powersat design is similar in size to Boeing's. It features use of mirror reflectors to concentrate sunlight onto its solar arrays, which are of gallium arsenide. (Courtesy Rockwell International Corp.)

The receiving antenna would be elliptical, some five miles wide and seven and one half miles long. This somewhat romantic Boeing painting emphasizes that below the receiver panels, microwave intensities would be low enough for farming or grazing. Actually, a more likely location for rectennas may be strip-mined coal land or depleted oil fields, which already will have been long used for energy production. (Courtesy Boeing Aerospace Co.)

Conversion system to produce electricity by taking advantage of the temperature difference between warm surface waters and cold deep waters. Commercial-size systems would need heat exchangers with the surface area of 150 football fields and would lose most of their performance with the growth of a layer of marine slime only one one-hundredth of an inch thick. If the plant were shut down for even a few days, the heat exchangers would be ruined by being overgrown with another form of marine life—barnacles.

So we can expect that future years will see increasing attention paid to a solar power system which works around the clock, can be built quickly, and offers the prospect of competitive costs for its energy. This, to be sure, is the power satellite. Should it go forward, this more than anything else would spark a space program of truly large dimensions. It is the power satellite which appears as the best initiative leading to space colonization.

The concept of the power satellite sprang full-blown from the mind of one man: Peter Glaser, vice-president of the consulting firm Arthur D. Little, Inc. In 1968 Glaser proposed that it would become possible to place arrays of solar cells in geosynchronous orbit, the arrays miles in dimensions and weighing a hundred thousand tons. The resulting electricity would be converted into a focused beam of microwaves and sent to Earth. There, it would be aimed at a receiving antenna or rectenna, which would convert the microwave energy back to electricity.

If the power satellite is to be practical, it will be necessary to develop vast new space projects. Immense space freighters will be needed to carry equipment to orbit at very low cost. Large crews of space workers will be shuttled to orbit and will need sustaining systems. The art of building space structures will need major developments. Above all, the cost of solar cells will have to drop from the present $10 per watt to $.50 or even less.

The power satellite thus is a tall order, but if it is not yet a formal project, it still is already something more than a gleam in the eyes of its proponents. Early in 1976 the Office of Management and Budget requested that ERDA, the Energy Research and Development Administration, consider the powersat concept as part of its solar energy program. An ERDA Task Group reviewed the NASA work on powersats and recommended a three-year study program to answer key questions.

The result was a joint NASA-Department of Energy ''Concept Development and Evaluation Program,'' which got under way in 1977. Funded at $15.6 million (an amount later raised to $22.1 million), it was a three-year study effort with the announced goal ''to build confidence in the viability of SPS [powersats] as a promising energy technology, or, at as early a date as possible, clearly identify barriers to SPS.'' However, as early as 1977, key ERDA managers were noting that ''no obvious and clearly insurmountable problems have been identified by the ERDA Task Group.'' By

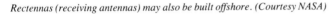

Rectennas (receiving antennas) may also be built offshore. (Courtesy NASA)

Rectenna panels would produce DC electricity, which would be converted to AC using solid-state electronics systems known as inverters and fed into the transcontinental grid of electric power lines. (Courtesy Boeing Aerospace Co.)

early 1979 the word was out: The study would favor the powersat and would recommend its study be funded as a promising new energy source.

More powersat funding is likely to come. Early in 1978 Congressman Ronnie G. Flippo introduced a bill to allocate $25 million to start technical development of powersats. (His motives were not exactly disinterested; NASA's Marshall Space Flight Center, a leading center for powersat studies, lies in his district.) In June 1978 the ''Flippo bill'' passed the House of Representatives by a vote of 267 to 96. A similar bill introduced in the Senate died in committee when Congress adjourned. Nevertheless, it was clear that Congress had shown its interest. The administration responded by requesting $8 million for powersat studies in fiscal year 1980, up from $6.6 million in 1979.

Passage of the Flippo bill, or else announcement of a new administration initiative following completion of the 1977–80 study effort would put the power satellite in roughly the same funding position as was fusion in the late 1950s. But the race is not always to the swift, nor to the earliest starter; the first commercial powersats could be on-line even before the first commercial fusion plants.

A competition between fusion plants and power satellites will be a most leisurely, drawn-out affair. Still, even before this competition is well begun, it is possible to make a small bet as to the winner.

It may be that fusion plants will have many similarities to the plants spawned by their parent technology, nuclear power. A fusion plant will probably be a huge, costly, complex affair, prone to expensive shutdowns. Its design will make it as much a plumber's nightmare as any nuclear plant. Its unreliability would not stem from the difficulty in operating numerous safety systems designed to keep its reactions under control; quite the contrary. The fusion reaction will be so difficult to start and maintain, so easily quenched, that it will take considerable effort merely to keep it running normally.

Fusion plants will not produce plutonium (at least, not without major modifications), and they will not be capable of a core meltdown. But they will still produce copious radioactive waste in the form of heavy reactor parts irradiated to the point where they lose strength and must be replaced, the replacement of which will be no mean feat. In addition, they will produce radioactive tritium, which mixes freely in water and will call for the strictest of controls. To a nuclear critic, it will not go unnoticed that the containment structure which physically houses a tokamak reactor will be indistinguishable from that which houses a nuclear one.*

---

*A tokamak must operate in vacuum; its containment building will keep the atmosphere out. This contrasts to that of a fission reactor, which serves as a safety measure to keep radioactivity in.

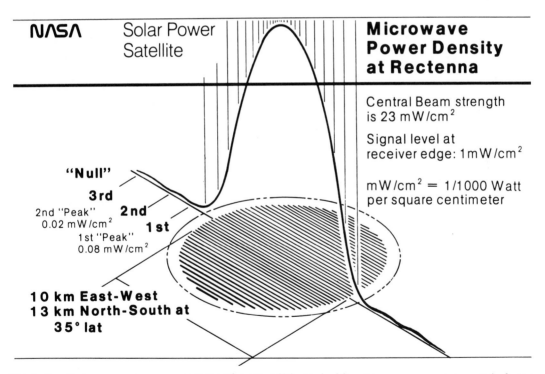

Distribution of microwave energy over a rectenna. The current U.S. standard for microwave exposure is ten watts/cm² which would be exceeded only in central regions of this fenced-off facility. Intense microwaves warm the body but have not been found to cause effects other than heating. (Courtesy NASA)

Orbiting powersats form a new constellation, the "Power Line." Actual powersats would be placed more randomly to better mimic ordinary stars and avoid disturbing the appearance of the night sky. (Courtesy Boeing Aerospace Co.)

Nuclear power has thus far been a technical disappointment, and its successor, fusion power, may share many of its weaknesses. By contrast, power satellites may flow from an area of technical strength: electronics. A powersat's solar cells will be solid-state electronic devices. Much the same will be true of the klystrons or microwave generators, which will transform the solar cells' electricity into the focused beam of microwaves directed to the rectenna. The control and shaping of the beam will be by other solid-state electronics known as ferrite-core phaseshifters. At the rectenna still other devices, the Schottky-barrier diodes, will convert the microwaves back to electricity.

The safety of a powersat will be assured without complex plumbing or opportunity for human error, for the powersat's transmitting antenna and the ground rectenna will have to cooperate. The transmitting antenna will have many small klystrons, all of which must oscillate in step like soldiers marching if the microwave beam is to be formed and focused. Soldiers cannot keep in step without a sergeant counting cadence; the "sergeant" will be the rectenna. It will send upward a pilot beam, fed with a small amount of its energy, which will serve as a reference signal. If the main power beam wanders, the pilot beam will go out, and with it the focus of that main beam. It will spread out, dissipating its power harmlessly into space.

The safety of microwaves has recently become a topic for research, but over forty years of experience with radar and similar enterprises have not shown any effect of microwaves other than warming, as in diathermy. Unlike radioactivity and its radiations, microwaves have not been found to damage cells or genes at low levels of exposure. Microwaves are much more nearly similar to the oscillating electromagnetic fields with which we all live night and day because of our household and office use of alternating electric current.

The present U.S. standard for exposure to microwaves is ten milliwatts per square centimeter, which is exceeded in the central regions of the rectenna. The rectenna thus will not quite be a place for family picnics. For protection of the general public, however, its most important safety feature may well be simple: a chain-link fence.

It thus is the powersat that can be a theme that will renew our reach into space, spark our hopes for the building of space colonies. There will be in this a taste of things to come. The colonies in time may serve as outposts from which we will face the vaster space that is the milieu of the stars. The powersats themselves will remind us of that, for from the ground they will be seen to reflect sunlight from their orbits, 22,300 miles up. Glistening, twinkling, shining in the night, the powersats themselves will appear as an arc of bright dots across the sky. They will be, indeed, new stars.

To build a powersat, however, will take more than electronics. There will be need for major advances in rockets. This, of course, is an old story; the dreams of astronautics have always flown as payloads requiring such rockets. Nor are such rockets new; their development and improvements have covered much of this century and today are still far from finished. Some people have suggested that the space shuttle will play a major role in building powersats, but in fact it will see only limited use. The actual rockets that will serve are the topic of the next chapter; and as usual, in keeping with the long-term character of astronautics, a bit of history will be in order.

CHAPTER 5

The Sweet Combination

T here is a very simple reason why the space shuttle cannot build powersats: It carries only thirty-two tons of cargo. A powersat thus would need three thousand shuttle launches just to haul its basic structure to orbit, to say nothing of the extra weight for its assembly facilities and for the propellant to take it to geosynchronous orbit. At sixty launches per year, the most the shuttle's launch facilities are planned to support, each powersat would take fifty years to build. Powersat enthusiasts thus can be forgiven if they seek something which will do the job more quickly.

Through the years many proposals and not a few projects have sought to achieve such advanced rockets. As early as 1898, the Russian pioneer Tsiolkovsky was writing that rockets for space flight should burn liquid fuel. In 1903 he pointed out the advantages of using hydrogen and oxygen as rocket fuels. In the days when wood-and-fabric biplanes were the very model of modern aircraft, not only Tsiolkovsky but also his American and German counterparts, Robert Goddard and Hermann Oberth, had predicted the performance of many of today's well-proven combinations of rocket fuels.

In the fall of 1945 rocket research began at NASA's Lewis Research Center, near Cleveland, now a leading center for research in advanced rockets. Overnight, the NACA (National Advisory Committee for Aeronautics, the forerunner of NASA) management switched the laboratory emphasis from piston engines to jets and reorganized the staff from top to bottom. A small group was even assigned to rocket research. The reorganization caught the lower-level supervisors and researchers by surprise. One scientist went home deeply engaged in writing a report on spark-plug fouling. The next morning he found his desk in another building, and he was officially engaged in research on rocket motor cooling.

The political climate in Washington was such that the directors of NACA did not want to proclaim that they were supporting research in anything as far-out as rockets. In Pasadena, California, this attitude dictated that a Caltech center for rocket research be given the name it holds to this day: the Jet Propulsion Laboratory. At NACA-Lewis the rocket group was officially called the High Pressure Combustion Section. Not till 1949 could the group emerge from cover and change its name to the Rocket Research Branch.

Even in 1945 rocket research was not new. The German V-2 project had already seen its wartime misuses. Its alcohol and liquid oxygen propellants, the standard rocket fuels of the 1940s, also fueled the Navy's Viking research rocket and the X-1 and X-1A rocket airplanes. The Jet Propulsion Laboratory was proving out such combinations as aniline and nitric acid; other laboratories were deeply involved in work on kerosene as a rocket fuel and in improving the performance of solid propellants. The Lewis people actually were rather late to the field, and to make a contribution they had to work in less plowed areas. They concentrated on liquid propellants of high energy.

Their propellant work was straightforward. They first computed the theoretical performance of candidate combinations, then selected the most promising for experiments. At times their evaluation led into more detailed studies of propellant characteristics, including the starting of rocket motors, their cooling, and problems of combustion. In 1948 Vearl Huff and his associates, who did this theoretical work, made a major contribution. They developed a rapid mathematical procedure for the calculations, which had previously been quite laborious.

Paul Ordin headed the early experimental work. He and his group investigated hydrazine, diborane, and ammonia, to be burned with liquid oxygen, hydrogen peroxide, or chlorine trifluoride. The rocket men were not averse to taking risks. One man was given a sample of hydrazine in another city and had to transport it back to Cleveland. The stability of hydrazine was in question, and it could not readily be shipped. He solved the problem by putting the sample in his pocket and bringing it home on the train.

In those days some important rocket propellants could be gotten only at one place, the Buffalo Electro-Chemical Company, and safety in transport was occasionally low on the list of priorities. When the first Viking rocket was in preparation for launch, in March 1949, the New Mexico launch crew found themselves running low on peroxide. Two of the Viking designers, in Baltimore and itching to be at the launch, solved the problem by driving a company station wagon to Buffalo. They put a fifty-gallon drum of the explosive fluid in the back seat, then drove at full speed despite snowstorms and other delays to reach New Mexico in four days—without benefit of interstate highways. Something like this happened at Lewis, too. Their first diborane came from Buffalo also and was delivered by an engineer in his own car. It came nestled in dry ice on the rear seat, complete with a safety device which soon saw use: a whisk broom. When some diborane leaked past a valve and spontaneously ignited, the engineer neatly whisked the flame away.

In May 1948 the Lewis people held a conference on their latest fuels research. The rocket research had used diborane with both liquid oxygen and hydrogen peroxide and had discovered a problem. Boron tended to form gluey combustion products, which produced deposits on parts of the motors. So it was that the rocket group became intrigued with the ultimate oxidizer, the most powerful available: liquid fluorine.

Other investigators had tested the use of gaseous fluorine, but Paul Ordin wanted to use it in liquid form and he succeeded. Initially he used diborane as a fuel. Huff's calculations had predicted high performance and no deposits of boron fluorides. However, the 9,200°F combustion temperature

The post-World War II Navaho missile was the focus for much early work in rocket propulsion. (Courtesy Rockwell International Corp.)

melted test engines in less than a second. What's more, they saw no reasonable way to cool a diborane-fluorine engine, for diborane is not a good coolant and fluorine is too reactive.

Their first experiments with fluorine, however, only whetted their interest. They worked with fluorine throughout the fifties, using it neat with ammonia-hydrazine mixtures as well as with ammonia and hydrazine. Their biggest effort with fluorine came after 1955 and involved burning it with liquid hydrogen. They also started a program to burn hydrogen with oxygen. The first studies of hydrogen-oxygen rockets dated to Tsiolkovsky in 1903 and Goddard in 1909. By the end of the forties, hydrogen-fueled rockets had been tested at Ohio State University, JPL, and Aerojet Engineering Corp. Aerojet's work was the most advanced, featuring a hydrogen liquefaction plant and three-thousand-pound–thrust motor tested in 1949. The Lewis rocket group wanted to work with engines up to twenty-thousand-pound thrust and with longer burning times.

To carry this program forward, NACA provided Lewis with a new rocket-testing facility. It was capable of not only operating twenty-thousand-pound–thrust hydrogen-fluorine rocket motors, but also of removing the deadly hydrogen fluoride from the exhaust and muffling or silencing the engine noise. The facility began operating in 1956. Over fifty thousand gallons per minute of water completely muffled the noise and absorbed the hydrogen fluoride. The water flowed into a tank and was treated chemically, producing an inert white powder (calcium fluoride), which could be hauled away. As one of the key people in this effort wrote in later years, "We were ahead of the environmentalists."

As this work advanced, it received an extra boost from one of Lewis' directors, Abe Silverstein. Silverstein was very enthusiastic about the potential of hydrogen not only for rockets, but also for aircraft. With his strong backing, the rocket group built and tested lightweight hydrogen-fluorine and hydrogen-oxygen motors, cooled with their own hydrogen fuel, with thrusts of five thousand and twenty thousand pounds.

By 1959 they had achieved full success with a hydrogen-fluorine motor under simulated conditions of space operation; actual performance reached almost 100 percent of theoretical. They measured a key performance parameter, exhaust velocity, at over 15,400 feet per second, the highest ever attained by a chemical rocket up to that time.

But the difficulties of working with liquid fluorine had not been lost on Silverstein. Work on hydrogen-oxygen had proceeded in parallel with the work on fluorine, and when Silverstein saw a hydrogen-oxygen motor run, the sweetness of that fuel combination came through to him loud and clear. Everything worked smoothly and simply, and performance was high.

In 1958 NACA gave way to NASA, and Silverstein was called to Washington to be director of Space Flight Development. His first task involved the Saturn project. Saturn had started in August 1958, when the Advanced Research Projects Agency (ARPA) gave Wernher von Braun the task of developing a booster with 1.5 million pounds of thrust using available engines. With the advent of NASA, President Eisenhower transferred the Saturn project to the new agency. The early emphasis had been on the first stage of Saturn, and Silverstein was named to head a group that would prepare recommendations for its upper stages.

Late in 1959 the committee recommended the hydrogen-oxygen combination for the upper stages of Saturn. The upshot was Saturn's use of the first commercial hydrogen-oxygen motor: the RL-10, built by Pratt & Whitney of Hartford, Connecticut. With fifteen thousand pounds of thrust, the RL-10 was the first hydrogen-fueled rocket to be built for actual use in space. It saw only limited use with the Saturn program, but it continues in use to this day with the high-energy stage known as Centaur.

Centaur launched the Pioneer and Voyager spacecraft to Jupiter and Saturn, as well as *Mariner 9* and the 1976 Viking missions to Mars.

The advent of the RL-10 meant that leadership in developing advanced rockets was passing into the hands of industry. Pratt & Whitney had grabbed the first plum, but it was not long before there were other key projects. Most of those went to an established industry leader: Rocketdyne, a division of North American Aviation.

North American got started in this area in 1946 with an Air Force contract to develop what today would be called a cruise missile. With the name Navaho, the project called for a ramjet-powered pilotless aircraft to fly 5,000 miles at three times the speed of sound at an altitude of 100,000 feet. Since ramjets develop thrust only when they are at high speed, the Navaho required a rocket booster to carry it to the speed and altitude where the ramjets could take over.

Even today, developing such a craft would be no mean feat; in 1946, the needed technology simply did not exist. The need was for new guidance systems, new electronics; for ramjets, for supersonic aerodynamics, and for rocket engines. To push forward in these areas, North American set up their Aerophysics Laboratory, with a rocket test facility in the Santa Susana mountains north of Los Angeles. In 1950 they began testing Navaho rocket motors there. With a company reorganization in 1955, Rocketdyne, headquartered in Canoga Park, was established as a separate division.

The nation's missile programs in the 1950s brought much new work on rocket propulsion. Rocketdyne won the early contract for the 78,000-pound thrust motor for the Redstone, America's first major missile, which burned alcohol and liquid oxygen. By the mid-1950s, Rocketdyne had contracted to develop a 150,000-pound thrust engine, burning kerosene and liquid oxygen. This engine, or slight modifications of it, was selected for four of the nation's most important rocket programs: Atlas, Thor, Jupiter, and Saturn.

As early as 1956 NACA-Lewis and Rocketdyne had started work on what would later become the F-1: a single rocket motor burning kerosene and oxygen, developing 1.5 million pounds of thrust. The early work concentrated on a forty-inch-wide injector for the propellants of this immense motor. In January 1959 the F-1 was established as a formal project to be built by Rocketdyne. In 1959 Rocketdyne engineers at Santa Susana were already testing an early version, which featured the forty-inch injector and a combustion chamber, though the chamber was built of heavy metal and was not cooled. The early test runs went only one and a half seconds. By June 1961 a complete F-1 was fired successfully. That was less than three weeks after President Kennedy committed the nation to achieve a lunar landing by 1970. When Kennedy announced that decision, he already knew that the F-1, key to the lunar rocket, would soon pass a critical test.

The F-1 project began fully three years before the design was chosen for the rocket that would use it: the Saturn V first stage. The NASA leaders knew that an early start on new rocket motors was essential; their decisions were to develop powerful new engines, without necessarily having a commitment to specific boosters that would use them. In this way, they laid much of the groundwork for the Apollo program even before Kennedy took office, in the days when Eisenhower was pooh-poohing suggestions that the nation expand its space initiatives.

The F-1 was not the only new motor begun in this way. In the summer of 1960 Rocketdyne won another contract: to develop a 230,000-pound thrust engine burning hydrogen and oxygen, to be known as the J-2. As with the F-1, the J-2 was begun well in advance of the space vehicles that would use it. The first successful ground test of a J-2 came in late January 1962. At about the same time preliminary designs were developed for the second and third stages of the Saturn V. The second stage,

The advent of hydrogen as a propellant brought a great increase in the lift capacity of America's rockets. Top row, left to right: *X-15, Mercury capsule, Gemini (with spacewalking astronaut), Apollo/Lunar Module.* Middle row: *Skylab, Apollo-Soyuz, shuttle orbiter with Spacelab.* Bottom row: *Atlas (Mercury), Titan (Gemini), Saturn V (Apollo), Saturn V (Skylab), Saturn I-B (Apollo-Soyuz), space shuttle. All of America's manned spacecraft since Gemini have used hydrogen. (Courtesy Rockwell International Corp.)*

designated S-II, used five J-2 motors, much as the first stage used five of the more powerful F-1's. The third stage, the S-IV B, used a single J-2.

This S-IV B functioned well in its first flight late in February 1966. A second flight, in July, disclosed no major problems. The first flight of the complete Saturn V took place on November 9, 1967, and again all was successful. But on the second test flight of Saturn V, in April 1968, a problem arose which threatened to bring the progress of the Apollo program to a dead halt.

The first stage functioned successfully, and the second-stage cluster of five J-2's started properly and at first operated normally. After 70 seconds, the engine compartment in the area of the No. 2 engine began to chill. This chilling was followed by a slight reduction in engine performance; its combustion chamber pressure fell off. Then, 193 seconds later, the No. 2 engine lost all pressure and shut down. This shutdown caused the adjacent No. 3 engine to shut down also. The onboard computer adjusted for the loss of these two engines by computing a modified flight path and longer burn time for the surviving three engines, and the spacecraft reached orbit successfully.

The single J-2 in the S-IV B third stage started normally and operated for 68 seconds. Then its engine compartment also began to chill, followed after another 40 seconds by a slight falloff in engine chamber pressure. The engine functioned in this fashion for the full 170 seconds of its planned burn. But when the engine was commanded to restart, it could not be made to do so.

This meant that despite its earlier successes, the J-2 could not be relied on for manned flight. Yet it was critically important that the problem be found and fast. Amid the euphoria of a string of successes in 1966, NASA planners had set the first lunar landing for February 1968. The tragic Apollo fire in January 1967 cost the lives of three astronauts and left NASA deeply shaken; many key managers were transferred or fired outright. That fire also cost the program a year and a half of delay, as the Apollo command module had to be redesigned. Now, the new J-2 problem threatened further delay.

Yet such delay was quite intolerable. The Apollo program had from its start held the goal: Beat the Russians to the Moon. By the spring of 1968 the achievement of this goal was in doubt. The Soviets were preparing their Zond spacecraft to carry a cosmonaut on a looping flight around the Moon. Such a flight would have been far short of a lunar landing or even of orbiting the Moon. Indeed, when the U.S. undertook such a single-loop circumnavigation, in the *Apollo 13* mission, it was only to save the astronauts' lives; the flight was written off as a failure. But a successful Soviet flight would have given them the highly prized achievement of the first man to the Moon. Later Apollo achievements—first to orbit the Moon, first to land on it—then would have been merely qualifications.

The threat to the J-2 was real. Extensive ground testing had failed to turn up situations that would duplicate the failures which had occurred in space. There was no way to recover the failed engines orbiting the Earth, and only the most meager data was sent back from onboard instruments.

Rocketdyne's J-2 project manager, Paul Castenholz, had the responsibility for solving the problem. Normally an intense, hard-driving man, Castenholz drove himself to new heights of intensity as Wernher von Braun called for a round-the-clock, seven-day-a-week effort.

The eventual solution turned out to be elusive but simple. Liquid hydrogen flowing through an auxiliary fuel line had set up a vibration, causing the line to rupture. In turn the supercold liquid was released causing the temperature drops in the engine casings. The failure had been missed in earlier ground tests because they had been conducted in normal atmosphere. The frigid liquid hydrogen had

Early space shuttle designs called for a two-stage fully reusable craft, with twelve engines in the booster, two in the orbiter. Engines were to be of the same type but with different nozzles. (Courtesy Rocketdyne)

actually caused air to solidify around the fuel line and to form an icy sludge that protected it from vibration. In airless space, there was no such protection.

To track through the few clues to the solution took thirty days. Thanks to the ingenuity and hard work of Castenholz and his group, Apollo went forward on schedule. In the wake of this success, NASA officials announced that the first manned Apollo flight using the Saturn V would orbit the Moon. This was the famous *Apollo 8* Christmas flight, whose astronauts read from Genesis and televised views of the Earth and Moon to the world. As for the Soviets, their unmanned test craft *Zond 5* successfully circumnavigated the Moon, but entered Earth's atmosphere along a wrong trajectory and experienced re-entry forces sufficient to kill a man. The Soviets then abandoned their effort.

Even before Neil Armstrong's "one small step" in July 1969, NASA planners had begun to look beyond the Apollo program to the project that was to become the space shuttle. Here again, advanced rocket development was pointing the way. As early as 1965 the Air Force has given a contract to Pratt & Whitney to develop a new experimental motor, the XLR-129. Burning hydrogen and oxygen and developing 250,000 pounds thrust, the XLR-129 incorporated important advances over the J-2. It was to operate at the high combustion chamber pressure of 3,000 psi, to produce high thrust and improved performance from a lightweight, compact engine. Also, it was to be reusable.

Preliminary studies of space-shuttle designs began in earnest in 1968. Many of the early designs called for direct use of the XLR-129 or of closely similar engines. By 1970 NASA requirements had

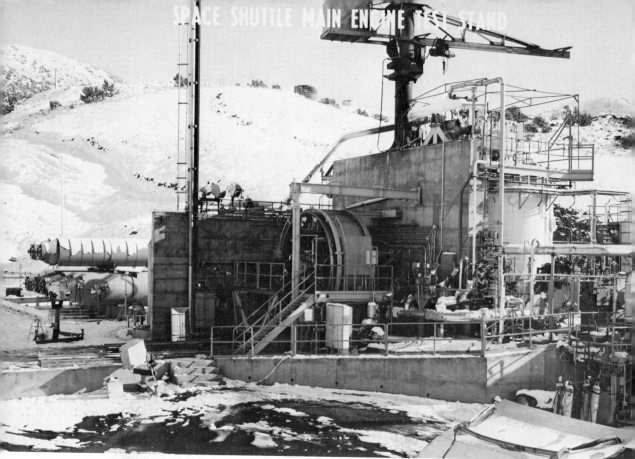

Test stand near Reno, Nevada, where initial experimental versions of the Space Shuttle Main Engine (SSME) first were put through their paces. (Courtesy Rocketdyne)

jelled, and called for a two-stage shuttle, each stage of which was to burn hydrogen-oxygen. Both stages were to be designed as winged rocket airplanes, flying back from their missions for frequent reuse; and both stages were to use what was by then designated the Space Shuttle Main Engine, or SSME. The SSME was to develop 550,000 pounds thrust, with twelve engines in the first stage, two in the second stage or orbiter.

Detailed studies of such shuttle designs went forward at North American, General Dynamics, and McDonnell Douglas during 1970–71. At the same time other detailed studies were under way on the SSME at Rocketdyne, Aerojet, and Pratt & Whitney. Most observers gave the edge to Pratt & Whitney, whose XLR-129, already successfully tested, gave them a decided leg up. At Rocketdyne, they had their hands full with existing contracts. It fell to Paul Castenholz, fresh from his J-2 triumph, to grasp the nettle.

He persuaded his corporate management to grant him $3 million of company funds to undertake the building of an experimental motor. His motor would go the XLR-129 one better since it would develop 400,000 pounds thrust, to the 250,000 of the competition. With his usual hard-driving approach, he set his group to work in December 1970. Two months later, he had his motor.

It featured a copper combustion chamber cooled with hydrogen in the manner of a space-rated engine and incorporated an injector design similar to the one that would finally fly. It also used preburners, as would the final design, in an important demonstration of SSME technology. In February 1971 the motor was taken to the Rocketdyne facility at Reno, Nevada, rigged to a test stand, and fired for one-half second—enough to measure performance, yet not so much as to risk burning a hole in the side.

This test completed, Castenholz had the motor taken on a tour of NASA centers. At Lewis, Marshall, and NASA Headquarters in Washington, NASA people could see for themselves that an SSME was just around the corner. That effort of Castenholz put Rocketdyne back in the race for the SSME contract; in July 1971 Rocketdyne won the contract.

In the summer and fall of 1971, as the restrictions of future tight budgets became evident, NASA completely changed the shuttle design. In a matter of months, it evolved from being a two-stage fully reusable design to the current design, which uses solid boosters and carries its propellants in an expendable external tank. These changes influenced the SSME. When the shuttle was redesigned, NASA proposed that it use not two but four engines in the orbiter. But the shuttle was to be built for frequent reuse, and it was expected to operate in the fashion of an airliner. Rocketdyne sought the

To win the SSME contract, Rocketdyne's Paul Castenholz built and tested this experimental engine using $3 million of company money. Exhaust flame is light blue and nearly transparent, a characteristic of hydrogen rockets. (Courtesy Rocketdyne)

counsel of TWA, whose consultants pointed out that they had experienced great savings in bringing in such three-engine airliners as the Boeing 727, as compared with four-engine craft like the 707. With fewer engines, aircraft required less maintenance. As a result, Castenholz recommended that the shuttle have not four but three engines, and the shuttle has remained so to this day.

The decision to build a three-engine shuttle fixed the SSME thrust at 470,000 pounds. The first complete engine was tested in May 1974; again the test was for one-half second. Then, over most of the next four years, the SSME program struggled through one major problem after another.

The problems involved the propellant turbopumps, which were driven by preburners. In the preburners hydrogen was burned with a minimal flow of oxygen, producing a very fuel-rich exhaust that was cool enough not to destroy the turbopumps. These gases drove the turbines, then fed through the injector into the combustion chamber, where they were burned with the rest of the oxygen, producing temperatures of 6,000°F. Hydrogen, flowing in many small tubes close to the main lining of the engine, cooled the combustion chamber and nozzle before being pumped to the preburners.

The fuel turbopump was designed as a compact package, about four feet long by eighteen inches in diameter, the size of a large outboard motor. It was to develop 76,000 horsepower. This was more than had run such huge ocean liners as the *Mauretania* early in the century in an era when ship engine rooms covered an acre and more of space below decks. Few comparisons can better illustrate the astounding advances in engine design that have occurred in this century. Of course, marine engines have been built to last for decades of service; the SSME was designed for 7.5 hours of total operation. Yet even this was a huge advance over the few tens of minutes of operation required of earlier rockets, which were only to be flown once.*

The first problem with the turbopump was that the turbine shaft was not held solidly enough by its bearings; it tended to jiggle with a circular motion. At 37,000 rpm, the jiggling quickly wore out the bearings. The solution was to stiffen them but it took eight months to figure out how to do it properly.

Next, there was the problem of cooling and lubricating the bearings. No oil could be used; thus the bearings were cooled with liquid hydrogen. Since this method was rather like lubricating an auto engine with water, it took some ingenuity. It turned out that one set of bearings was not getting enough hydrogen, was overheating, and then failing. The solution: redesign the channels that fed the bearing its hydrogen. That took another six months.

A third major problem involved the turbine blades. There were sixty-three of them in one section of the turbine, each the size of a postage stamp; and each of them was generating six hundred horsepower, the power of an Indianapolis racing car. Naturally, the blades were under some strain, and they tended to crack. The problem was traced to vibration, a well-known scourge of engineers, and finally corrected. The time: another six months.

While work was progressing (slowly) on the fuel turbopump, there was concurrent development of the oxygen turbopump. Here too there were serious problems but the problems were more difficult to find and correct. When turbopump problems made their presence felt, the turbopump in question would fail and shut down. A fuel turbopump failure simply caused the engine to lose power. It could then be disassembled, and the cause of the failure sought. But failures in the oxygen turbopump

---

*In this, there is considerable hope for the future of low-cost space flight. One of the early jet engines, the J-33 used in the F-80 fighter, in 1946 had an average operating lifetime of fourteen to twenty-five hours. Modern jet engines have lives up to 10,000 hours. Since engines are expensive, long life helps greatly in keeping costs down. Similar improvements in rocket engine lifetimes may be in store in decades to come.

brought severe damage to the engine. In earlier rocket development projects, oxygen pump failures often caused the motor to blow up. The SSME was built of sterner stuff; to contain its high pressure, it was built so stoutly that it would not (usually) explode. But it could, and did, catch fire. It was built of metals like copper, nickel, and steel, which we do not regard as fire hazards; but at the temperatures and pressures of an SSME, in the presence of liquid oxygen, virtually anything will burn. These fires often burned so much of the engine that it was difficult to discover what parts had failed or in what sequence the failure had spread.

The oxygen turbopump suffered two major problems. The easy one took six months to find and fix. It involved a rotating seal, which served to separate liquid oxygen from the hot gases in the turbine. The seal was designed to rotate without friction, but it tended to rub against another engine part. This rubbing then produced heat by friction—heat sufficient to ignite the metal, as when Boy Scouts start a fire by rubbing sticks together. In the SSME such fires burned more than marshmallows. Eventually, by choosing a different type of seal, the problem was licked.

The most difficult problem, though, was that the oxygen turbopump bearings repeatedly failed and burned up. In the end, a variety of expedients was used. The rotating shaft of the turbine was redesigned to give better balance. Just as an unbalanced auto tire wears rapidly, the inadequately balanced turbine shaft, rotating at 31,000 rpm, had worn so quickly as to cause the unit to fail. The bearing supports were stiffened. Finally, the bearings and their races, or holders, were made bigger and built to carry heavier loads. After a year and a half, the turbopump bearings finally passed their tests.

The last pump bearing failure was in March 1977, and from that point the designers could proceed with engine tests at low power. They could not proceed to tests at the rated power level till they were quite sure that there were no problems at the lower levels, and the testing proceeded cautiously. By mid-August 1978, engine tests had accumulated a total of only 17,000 seconds of operating time, much of which had been at low power.

Late in that month, Rocketdyne officials met with NASA administrator Robert Frosch and stated that they were prepared for a dramatic step-up in their rate of testing. The goal to be reached by September 1979 was 65,000 seconds of test time, at which point the engine would be pronounced qualified for shuttle flight. The Rocketdyne people were as good as their word, and better. By Thanksgiving they had reached 32,000 seconds, and most of the new testing was at 90 percent or more of rated power. The accomplishment put them three-and-a-half months ahead of schedule. Thus, they set themselves a new goal: to reach the 65,000-second mark by early June. That would qualify the SSME for a first shuttle flight date of September 28, 1979. Very soon the testing was a month ahead of even the new schedule.

Then just after Christmas 1978, disaster struck again. An SSME motor, under test at the major NASA facilities at Bay St. Louis, Mississippi, blew up. This time it was not a problem with the turbopumps; the new problem areas were the main oxidizer valve and a heat exchanger. The valve problem, once realized, was quickly fixed by a redesign. The heat exchanger problem was less simple. In the words of a senior manager, ''The failure of the heat exchanger remains unexplained, and it gives you a very soggy feeling. These incidents occurring so late in the test program just do not inspire confidence.''

Further testing was held up for weeks, and Frosch slipped the first shuttle flight to November 9, 1979. As 1979 progressed, continued SSME testing proceeded, and no new problems were uncovered. Confidence in the engine again began to grow.

Today's SSME reflects the harsh lessons of fires and explosions on the test stand, and of other failures. (Courtesy Rocketdyne)

Preparations at Cape Canaveral for launch of the first space shuttle, expected in 1980. (Courtesy Rockwell International Corp)

The same confidence did not extend to the November 9 launch date. By June 1979, it was evident that first launch would slip at least to June of 1980. At the same time, NASA officials announced new delays in the building of the orbiter craft and stated that the overall shuttle program would come in with a billion-dollar cost overrun, when compared with the original budget plan set in 1972.

Why these delays and overruns? In the views of senior program officials, the reason is that the program had been too thinly supported from the start, its funding levels held too low to assure success on time. With more funding, there would have been better analysis and testing of designs, which would have uncovered flaws earlier, allowing them to be corrected more easily. As it was, the program was managed with a strong "success orientation": it lacked the sort of managerial defense-in-depth which could readily accept failures and errors, then work to correct them. Such technical surprises were only to be expected in view of the unprecedented requirements of the shuttle program, and in the equally demanding Apollo program, it was just this defense-in-depth which Castenholz had relied on in rescuing the J-2 engine. Because of this, Apollo had reached the Moon on time and within budget. But by skimping on funding early in the shuttle program, problems had been left to build up, until they could be fixed only at considerable expense and delay.

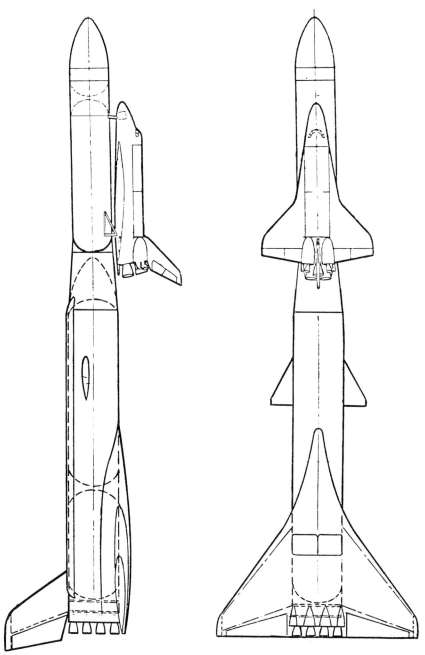

Shuttle-Derived Vehicle (SDV) proposed by General Dynamics would extend the usefulness of the shuttle orbiter. Its flyback first stage, burning propane and oxygen, closely resembles the booster of the early space shuttle design, which President Nixon in 1971 decided not to build. (Courtesy General Dynamics)

Single-stage-to-orbit (SSTO) launch vehicle would use Robert Salkeld's principle of mixed-mode propulsion for high performance. (Painting by Bill DeGeneres. © 1974 System Development Corp. Courtesy Robert Salkeld)

For all this, there was no doubt the shuttle soon would fly; and by mid-1979, the SSME had long since shown the true meaning of the phrase, ''engineering development.'' The long and weary years of turbopump shutdowns and engine fires were now a receding memory. Meanwhile, interest in power satellites was steadily increasing. In short, it was possible to look ahead to a new, major rocket engine development project.

In reality, rocket specialists had never entirely ceased their thinking about the future, and such projections had gained great encouragement in 1971. In that year Robert Salkeld, a consultant with Systems Development Corporation, announced his discovery of mixed-mode propulsion. His discovery raised eyebrows among rocket experts, for he had found a new wrinkle in the very familiar equations which govern the performance of rockets. First derived around the turn of the century, these equations are taught in freshman physics classes, and have been worked over so thoroughly that it was difficult to believe anyone could find anything new in them.

Salkeld had studied the performance of a rocket stage carrying two different fuels and two types

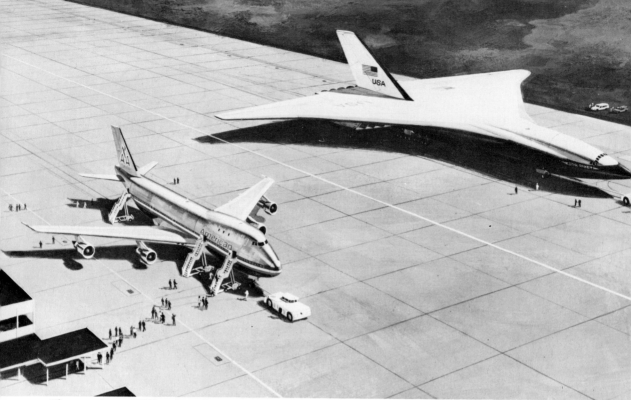

The most advanced SSTO craft now foreseen would use scramjets for propulsion. Here one such craft Star-Raker *prepares for takeoff. An old but still serviceable Boeing 747 is in the foreground. (Courtesy Rockwell International Corp.)*

of rocket motors, each burning its own fuel. He showed that a judicious choice of the fuels and of the sequence in which they were to burn would allow a rocket stage of given size to carry more payload than with any single choice of fuel, even if the fuel was of high energy like hydrogen.

For instance, a space shuttle might be designed as a single stage, burning only hydrogen and oxygen. But if part of the fuel tank was given over to a different fuel such as kerosene, then the shuttle could be made to carry much more cargo, or else the craft could be built smaller, more compactly, and still carry the same payload.

The obvious application of this idea was to a long-held dream of rocket designers, the single-stage-to-orbit rocket, or SSTO. Such a craft would take off (possibly from a runway, in the fashion of an airplane), fly to orbit in one piece and perform a mission, then return through the atmosphere to a landing. It would bring true airplane-like simplicity to space operations by avoiding the costs and complexities of booster stages. And it would bring a certain tidiness. After all, every airplane that flies today is a single-stage craft.*

Salkeld went on to work with Rudi Beichel, an expert in the design of rocket motors, to propose SSTO craft capable of a most diverse range of missions: space rescue, cargo transport to orbit, even transport of people by rocket to any point of the world in two hours. An important key to these ideas

---

*This was not always so. In the late 1930s the task of commercial flight across the Atlantic was thought by some to be so difficult as to require a two-stage aircraft. This, the Short-Mayo Composite of 1938, was a four-engine flying boat, which took off with a smaller four-engine seaplane on its back. It worked. But by 1939, the larger Boeing 314 had introduced single-stage transatlantic service.

was a novel rocket motor designed by Beichel, the dual-fuel motor. It would start off at liftoff burning kerosene or a similar fuel, but would later switch to hydrogen.

In 1975 NASA undertook a study, "Outlook for Space, 1980–2000." An attempt to forecast the nation's future in space, the study called on the nation's best rocket designers to come forth with their predictions as to what rockets of the future would be able to do. In particular, they called for suggested designs for a class of rockets capable of use in building power satellites. Such rockets would carry two hundred tons to orbit. The resulting designs included both single-stage and two-stage craft, water landings and aircraft-type landings by winged craft. The SSME figured prominently in many of the designs, but the designers were not reticent in calling for new engines.

In 1977 two Langley scientists (Beverly Z. Henry and Charles Eldred) reviewed their center's work on advanced rockets in a paper given at the Princeton Conference on Space Manufacturing Facilities and Space Colonies. They were concerned with the advantages to be gained from a novel rocket motor design proposed by Beichel: the dual-expander engine. This engine would make use of Robert Salkeld's principle of mixed-mode propulsion to offer performance gains of up to 25 percent over the SSME.

The dual-expander engine is actually two rocket motors in one. An inner combustion chamber burns oxygen and a fuel such as kerosene at the very high pressure of 6,000 pounds per square inch—twice that of the SSME. Around the inner combustion chamber is an outer one, which surrounds it to form a ring. It burns hydrogen and oxygen at the SSME's pressure of 3,000 pounds. The general arrangement thus resembles a bell mounted inside a larger bell. An ingenious flexible seal or bellows permits the combustion chambers to shift up or down with respect to each other, despite the hot, high-pressure gases. It serves to vary the size of the rocket's throat or narrow part of its nozzle, so

To reach orbit, Star-Raker *carries auxiliary rocket motors. (Courtesy Rockwell International Corp.)*

as to give the best performance at any altitude. An important feature is that at any time it is possible to thrust with either chamber, or with both, thus providing unparalleled flexibility in operation. Equally important, this high performance is achieved with a much shorter and simpler rocket nozzle than had previously been thought possible. The short nozzle makes it easier to mount in a rocket and easier to steer from side to side when controlling the rocket's flight. Such a motor can then produce the thrust of an SSME, yet have less than half the weight. A dual-expander motor weighing only 4,350 pounds could produce 608,000 pounds thrust at liftoff, compared with a 6,250-pound SSME, which produces 375,000. In the vacuum of space, where thrust increases, the SSME will give 470,000 pounds thrust, but the dual-expander design will give 693,000.

As Henry and Eldred emphasized, the dual-expander engine is far from the only advance that may be expected. Their paper looked ahead to continued progress in the design of lightweight structures: wing and tail surfaces, propellant tanks, engine mounts. They emphasized that modest levels of funding for research could produce large payoffs in reducing the weight of these structures. Other advances could be expected in protecting against the heat of atmosphere re-entry.

The lowest-cost rocket they discussed was a winged single-stage craft, which would take off

Two-stage space freighter proposed by Boeing would carry 600,000 pounds to orbit. (Courtesy Boeing Aerospace Co.)

Separation of first and second stage of Boeing's space freighter. First stage (at left) will be braked by parachutes and small rockets for a landing in the Atlantic. Second stage will fly on to orbit. (Courtesy Boeing Aerospace Co.)

vertically like the space shuttle, carry a 250-ton payload to orbit, then re-enter and land on a runway. With conventional SSME-type engines, it would be so tail-heavy that it could not fly in the atmosphere. It would lack the stability needed for safe operation. The lightweight dual-expander engines would overcome this problem by reducing the tail-heaviness. The result would be a space transport craft offering a cost to orbit of $7.20 per pound.

Still, such a craft would have to operate from Cape Canaveral or a similar site; it could never fly from such airports as JFK or Dallas-Ft. Worth. Precisely this type of operation would be possible with an entirely different type of propulsion—the scramjet or supersonic combustion ramjet. A type of jet engine, it would permit an aircraft to fly at speeds and altitudes that today can be reached only by powerful rockets.

Can a jet engine fly at close to orbital speeds? A turbojet cannot fly at much more than three times the speed of sound, or as an aerodynamicist would say, Mach 3—no mean achievement. The reason for this performance is that incoming air must be slowed down and compressed in order to burn the fuel by means of a carefully shaped inlet channel that creates shock waves. Air passing through the shocks slows and compresses. In doing so, it also heats up, the heating growing intense at high speeds. Above Mach 3, the heat is sufficient to cause engine parts to soften or weaken. Nor can the engines be built to use cooler air flowing at higher speeds; the rotating fan or compressor of a jet then could not work properly.

Since the end of World War II propulsion specialists have been tantalized by the ramjet. The simplest type of jet in current use, it flies at high speed and rams or compresses air by its speed alone. It has none of the complex turbines and compressors of a conventional jet. It cannot take off from the ground, but requires a turbojet or rocket to get up to the several hundred miles per hour needed for it to work properly. Its simplicity allows it to fly higher and faster than a turbojet; so far from being limited to Mach 3, it reaches peak performance at that speed. It then can fly on to higher speeds called hypersonic. However, above Mach 6 its performance falls off drastically. By contrast, orbital velocity is Mach 24.

The reason is that the airflow in a ramjet still must be subsonic in order for the flame not to blow out. Again it is shock waves that serve to slow and compress the outside airflow to the desired internal speed. Though fierce heating is produced, a ramjet is simple enough that it can be cooled by flows of fuel. Above Mach 6, even this fails. The need for subsonic combustion sets the limits to performance of ramjets.

Since the late 1950s it has been appreciated that if a ramjet was capable of burning fuel in a supersonic airflow to provide thrust, very great gains in flight speed and performance would become possible. If the resulting scramjet would burn hydrogen, it would give better fuel economy at Mach 6 than a conventional turbojet at Mach 2. As early as 1965 scientists at the Marquardt Corporation operated an experimental scramjet at better than Mach 10. Since then, it has become a certainty that scramjets can be developed to permit routine flight well above Mach 6.

Then why are there no scramjets today? The reason is that they pose peculiar problems not encountered in developing jet engines and rockets. A rocket can be tested on the ground and its performance in space reliably predicted. The same is true when a turbojet is ground-tested or is fed with high-speed air from a wind tunnel. A scramjet can also be run in a wind tunnel, and indeed many have been. But the test results cannot reliably be used to understand how a scramjet would actually perform in hypersonic flight.

The reason is that a scramjet has a curious feature. Unlike a turbojet or rocket, it cannot be designed and built separately from its aircraft and then mounted. Instead, aircraft and scramjet must be designed together. The forward portion of the aircraft underbelly actually must behave somewhat as an inlet channel, which feeds air to the scramjet. The aft portion of the underbelly acts as a nozzle, allowing the scramjet exhaust to expand and produce more thrust.

Such a complex system cannot be tested as a scale model in a wind tunnel. Only a full-size craft will do, and only in hypersonic flight. In the mid-sixties NASA had a Hypersonic Research Engine project that aimed at building an experimental scramjet to fly aboard the X-15 rocket plane. However, the X-15 program was cancelled (in 1968) before the engine was ready to test. If scramjets are to be developed, there will be need for an entirely new Hypersonic Research Aircraft, which indeed is strongly advocated by the senior management of NASA-Langley. Like the earlier research aircraft such as the X-1, X-2, and X-15, this craft would continue the well-proven practice of pushing back the frontiers of flight by means of research aircraft flown by daring test pilots.

It took a quarter-century to advance from the first experimental flight at Mach 2 to the double-sonic Concorde commercial jet. It would probably take nearly as long to build and test this new research craft and to apply its lessons to a new generation of single-stage-to-orbit vehicles. Such scramjet craft would need still more time for development. So for the task of building powersats, scramjets will loom too far in the future to be of interest. Instead, the emphasis will remain fixed upon that tried and true engine, the rocket. It is this which will drive the powersat's space freighters.

Winged cargo rockets may resemble immense two-stage versions of the shuttle orbiter, particularly if they use the orbiter's wing planform. (Courtesy Boeing Aerospace Co.)

By 1978 further studies of power satellites had defined more precisely what the post-shuttle space freighters might look like. Such projected advances as mixed-mode propulsion, which made SSTO attractive, were seen to make two-stage freighter craft even more attractive. The single-stage craft offered the lower operating costs, though not by much; and two-stage craft still looked easier and less risky to design. With proposed payloads now approaching five hundred tons, the emphasis had to be on avoiding risk.

Even so, Salkeld's work had had its influence, and it was clear to all that the first stage of such two-stage freighters would have to burn a fuel such as liquid methane, or perhaps liquid propane. Thus it would be necessary to go forward with developing a new rocket motor.

This fact implies that advocates of the power satellite will have a clear signpost to look for, an indication that the nation truly is preparing to go ahead with building powersats. The new engine will have to be contracted for well in advance of any other major powersat activity. Like the F-1 and J-2, it

109

will likely start as a project and be well under way even before there is a commitment to go forward with the space freighter that will use it. Like the SSME and the earlier XLR-129, its development may be well advanced before firm designs are set for the freighter, or even for the powersat itself.

The sign to look for will be a NASA contract, probably to Rocketdyne, to proceed with a new and large rocket motor. Its thrust will be some two million pounds, perhaps more. It will burn oxygen and methane or propane and may be capable of dual-fuel operation. It will be designed for high chamber pressure and consequent compact size, and for operating lifetimes of many hours.

On the day that contract is announced, probably in the early 1980s, it will be evident to those who are knowledgeable that whatever the president may say, whatever that year's space budget may show, a commitment to build powersats will be coming along. It will come as certainly as the digging of a foundation presages the building of a house.

CHAPTER **6**

Large Space Structures

Imagine a small spacecraft carrying an inspection party has landed in the middle of the vastness of a powersat. From the observation ports, a few meters above the flat, level expanse of silicon cells, the powersat extends in all directions to the horizon, like a featureless, trackless stretch of desert.

Now the spacecraft begins to rise, to move away from the powersat like a grasshopper flying up from a beach. (The relative dimensions are rather similar.) For quite a while, there is little new or different to see. The world outside continues to be sharply bisected by what to all appearances is a plane extending to infinity. A while longer, a further elevation, and the plane is seen to have edges. It does not fill the universe, but its extent still cannot be grasped.

Only much later, when the distance has increased to tens of miles, can the true dimensions be seen. It is a flat rectangular slab, with circular disks at each end—and the whole of it is larger than Manhattan Island. It spans the area of 20,000 football fields. It has the weight of an aircraft carrier, yet could enclose the flight decks of 6,000 such carriers.

Is this the work of a super-civilization of the year 3000? The artifact of extraterrestrial visitors, commanding sciences and technologies of which we can only dream? No, it probably is the work of people who are alive today, and the year could be as early as 1995. And if a Manhattan-size spacecraft is hard to conceive, then consider an everyday item familiar to all, the production of which far outstrips the dimensions of powersats: the daily newspaper.

The Los Angeles *Times,* which I read every day, illustrates this point most concretely. The masthead carries the legend: "Circulation, 1,034,329; 152 pages." That is, between Monday and Thursday of each week, the *Times'* printing plants turn out enough newsprint, printed neatly on both sides, to more than cover the total 114 square kilometers of a complete power satellite.

Flex-Rib communications antenna used on the satellite ATS-6. (Courtesy Lockheed Missiles and Space Co.)

Small crews living in the space shuttle will carry out early space construction work. (Courtesy Rockwell International Corp.)

The proposals of astronautics may boggle the mind; but its history shows that time and again, even the most startling concepts have swiftly become commonplace, even prosaic. The Boeing 747 is a case in point. Another is the communications satellite, which only fifteen years ago was grist for the science fiction of Arthur C. Clarke. Twenty years ago an artist's rendering showing the space shuttle maneuvering near the space telescope could only have been wild fantasy; today such a picture is nothing more than an advertisement for Lockheed Missiles and Space Company. So, too, will it be with the powersat. Vast though it is, it lends itself to rapid building by construction techniques that will in time appear so straightforward as to be, indeed, prosaic. Sooner than we think, a new powersat may be no more remarkable than a new day's edition of the *Times*.

The initial goal of work on large space structures will of course be nothing so elaborate as a power satellite; it will rather be the building of large communications platforms of the type discussed in Chapter 3. But the people who are studying such platforms are well aware how the techniques needed to build them would serve as well on a larger scale to build powersats. These techniques, even in their simplest applications, are quite novel and represent a considerable advance over what has been done to date.

It is no new thing to build a satellite larger in dimensions than the payload compartment that

DEPLOYABLE ANTENNA EXPERIMENT

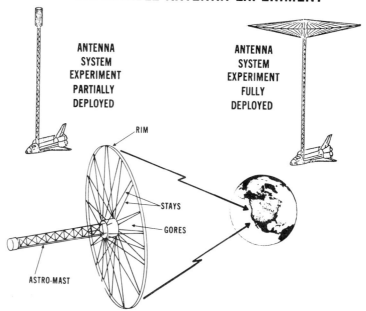

ANTENNA
SYSTEM
EXPERIMENT
PARTIALLY
DEPLOYED

ANTENNA
SYSTEM
EXPERIMENT
FULLY
DEPLOYED

RIM

STAYS

GORES

ASTRO-MAST

Earliest use of the shuttle in space construction will feature large deployable structures using the extendable Astromast. (Courtesy NASA)

Shuttle with large deployed antenna. (Courtesy Grumman Aerospace Corp.)

GRUMMAN

COMPOSITE BEAM BUILDER

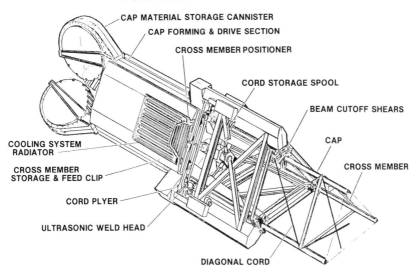

CAP MATERIAL STORAGE CANNISTER

CAP FORMING & DRIVE SECTION

CROSS MEMBER POSITIONER

CORD STORAGE SPOOL

BEAM CUTOFF SHEARS

CAP

CROSS MEMBER

COOLING SYSTEM RADIATOR

CROSS MEMBER STORAGE & FEED CLIP

CORD PLYER

ULTRASONIC WELD HEAD

DIAGONAL CORD

Beam-builder studied by General Dynamics uses composite materials such as fiber-reinforced plastics. (Courtesy NASA)

Beam-builder built at Grumman produces beams of aluminum. (Courtesy Grumman Aerospace Corp.)

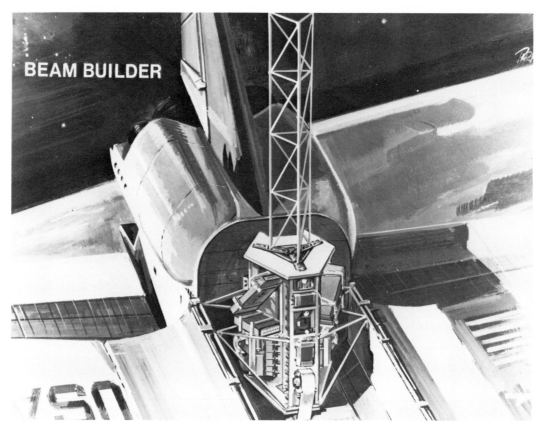

BEAM BUILDER

Beam-builder in the cargo bay of a shuttle orbiter goes through a test in Earth orbit. (Courtesy Marshall Space Flight Center)

carries it to orbit. There has long been ample need for creativity in designing spacecraft that unfold solar panels, deploy large antennas, or extend long struts once in orbit. In these areas examples of ingenious concepts abound. The FRUSA (Flexible Roll-Up Solar Array), for one, packages a spacecraft's power supply into something very much like a roll of linoleum. The Flex-Rib Reflector, used on the ATS-6 satellite, is a parabolic antenna whose form is shaped by forty-eight flexible ribs, such as those of an umbrella. The ribs are furled or wound round a central hub to form a lightweight, compact disc. Once in orbit, the ribs release and unfurl, standing out from the hub to form an accurately shaped antenna thirty feet in diameter. The Astromast is a long beam, which extends out to thirty times its packaged length.

In the era of the space shuttle, these approaches will appear as expedients, as ways of getting around the small payload bays and lack of on-orbit human assemblers. But within a few years, it will become routine to have work crews fashioning spacecraft structures on-orbit, working with a very basic and familiar structural element: the beam.

Even in the early 1970s the earliest studies of space structures showed that they would be built on-orbit with frameworks of beams. At first it was believed that the beams would be fabricated on the ground and carried to orbit by the shuttle. But such beams then would necessarily be limited to the sixty-foot length of the shuttle's cargo bay and in addition, would have to be narrow and compact, somewhat like aluminum pipes to take advantage of the shuttle's 65,000-pound payload capacity.

What the designers really wanted, though, would be beams of any length, up to hundreds of meters if need be. But the beams could not be narrow or they would bend easily. Instead, they had to be of a light, open design. Such beams could be produced by work crews in orbit, but an early Grumman Aerospace Corporation study of powersat construction projected an orbiting crew of 258, just to fabricate the beams. With this, it became clear that the beams would in no way be built on Earth. Instead, a new and daring approach was in order: What the shuttle (or its advanced successors) would carry to orbit would be rolls of aluminum, structural parts— and a new type of machine known as a beam-builder. This machine, operating in orbit, would then form and shape the aluminum into beams of the proper type, in what would amount to an orbiting construction plant.

STRUCTURE FABRICATION SYSTEM
CONCEPT FOR LADDER CONFIGURATION
LONGITUDINALS

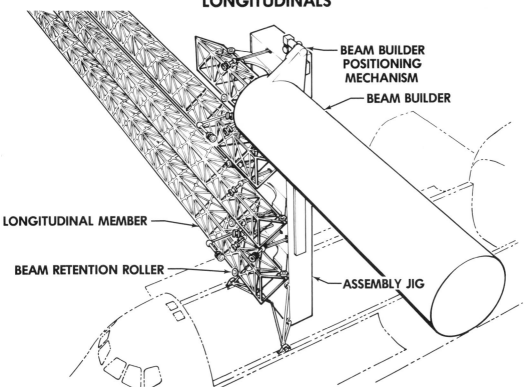

Use of a beam-builder and jig in the proposed Space Construction Automated Fabrication Experiment (SCAFE). (Courtesy Johnson Space Center)

Fabricating the crossbeams of SCAFE. (Courtesy General Dynamics)

PACKAGING AND DEPLOYMENT OF THE
CONSTRUCTION/DEMONSTRATION PLATFORM

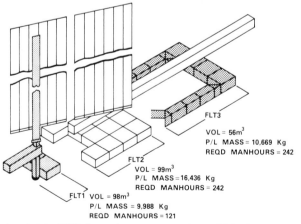

FLT3

VOL = 56m³
P/L MASS = 10,669 Kg
REQD MANHOURS = 242

FLT2

VOL = 99m³
P/L MASS = 16,436 Kg
REQD MANHOURS = 242

FLT1 VOL = 98m³
P/L MASS = 9,988 Kg
REQD MANHOURS = 121

Orbital Construction Demonstration Article (OCDA), used to support major activities in space construction, would be assembled from materials brought to orbit with three flights of the shuttle. (Courtesy NASA)

Completed OCDA would feature a solar array generating 250 kilowatts and a construction boom 360 feet long. (Courtesy NASA)

Boom of the OCDA would carry tracks for movable carriages. Such carriages would support small cranes, manipulators, cherry-picker cabins, and other construction equipment. (Courtesy NASA)

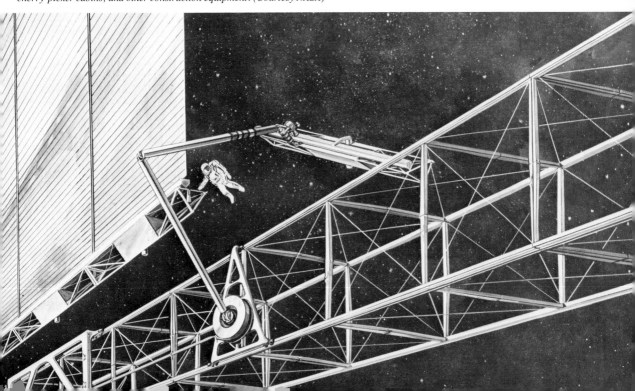

CONSTRUCTION OF RADIOMETER

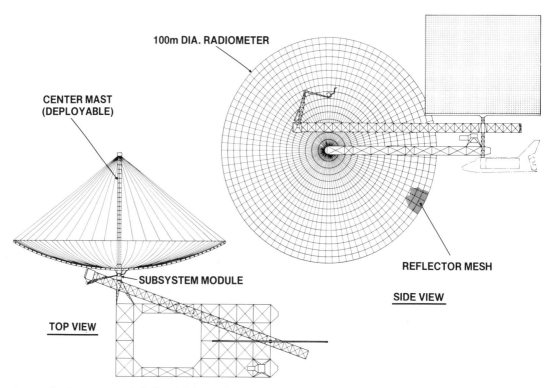

100m DIA. RADIOMETER

CENTER MAST
(DEPLOYABLE)

REFLECTOR MESH

SUBSYSTEM MODULE

SIDE VIEW

TOP VIEW

An example of a structure to be built with OCDA is this 100-meter radiometer for soil moisture measurements. OCDA would be operated by a crew in the shuttle orbiter. (Courtesy NASA)

So it was that NASA's Marshall Space Flight Center awarded a $635,000 contract to Grumman to build a "Space Fabrication Demonstration System"; that is, a beam-builder. The first such device was completed and delivered to Marshall in 1978. On May 4, 1978, it produced its first beam in ground test.

The beams it fabricates are both lightweight and strong. A one-hundred-foot length weighs only 85 pounds, yet will support a load of 1,260 pounds. The beams are triangular in cross section and a meter deep. (The depth of a triangular beam is the distance from one corner to the opposite side.) They are made up of long strips of angle aluminum supported by cross-braces. The long aluminum edge members are formed from rolls of sheet aluminum; the machine pulls out aluminum strips from the rolls and forms them into the proper angled shape. The cross-braces are made beforehand and packaged in magazines, which fit to the side of the beam builder. They are withdrawn automatically, somewhat like giant staples, and the machine automatically welds them to the edge members. With one supply of rolls of sheet aluminum and of full magazines of cross-braces, the machine can turn out a thousand feet of beam in as little as two hours.

The beam-builder will not long remain on Earth. As early as 1983 it may be modified for flight, placed in the shuttle's cargo bay, and put through its paces at an altitude of several hundred miles. This shuttle-beam builder system would do more than merely fabricate beams in orbit, though. The shuttle carries, as standard equipment, a long manipulator arm that is controlled by a crew member, the payload handling specialist, who works at the rear of the shuttle flight deck. This manipulator arm can act as a crane, wielding beams and partially assembled structures. A proposed 1984 flight test may demonstrate the full power of this system, known as the Space Construction Automated Fabrication Experiment (SCAFE), which has been studied by General Dynamics under contract to NASA's Johnson Space Center.

In SCAFE, the beam-builder and its raw materials would be carried to orbit along with a jig or assembly frame mounted to the shuttle. Once in orbit, the manipulator arm would serve to deploy the stowed equipment. Then the beam-builder, moving to successive positions along the jig, would automatically make four triangular beams, each 200 meters long and held parallel to each other by the jig. The beam-builder would then move to the outboard end of the jig and fabricate the first of nine short cross-beams, 10.6 meters long, spanning the four long beams. To attach these cross-beams at the proper sites, like rungs along a ladder, the jig would shift the partially completed beam assembly along its length, stopping every 25 meters to permit the beam-builder to fabricate another cross-beam.

Orbiter, OCDA, and completed radiometer. (Courtesy Grumman Aerospace Corp.)

Multibeam communications antenna, with 272 individual beams, would be built up from hexagonal cells using the OCDA. (Courtesy NASA)

The most exciting uses of OCDA will be in testing construction techniques for building power satellites. Here a twenty-meter-deep beam, such as will be used in powersat construction, takes shape beneath the OCDA platform. Eventually it will reach a length of 246 meters and then will be shaken in vibration tests. (Courtesy Grumman Aerospace Corp.)

Jig-mounted welders would join beam to beam where they cross. In this fashion a complete platform would be built, a grid of four long beams braced by nine cross beams.

A number of important experiments will then be possible. With the platform still firmly attached to the shuttle, the shuttle can fire the maneuvering engines and accelerate. This procedure will place a load on the structure, tend to bend it, and test its strength, particularly the strength of the welds made by the beam-builder. When the shuttle passes in and out of Earth's shadow, the structure will be subjected to sudden outside temperature changes and their effects can be measured. There can be tests with the manipulator arm, using it to release and recapture the platform, retrieving it as it floats freely in space. In addition, there can be tests of the performance of astronauts, who can attempt to accomplish such matters as attaching instruments and electronic equipment to the platform.

The shuttle, however, was never designed as a space construction center. Even with the aid of a beam-builder and of its manipulator arm, it can construct only modest structures. The shuttle has only limited power (seven kilowatts) from onboard sources; the manipulator arm cannot reach very far. To extend the capabilities of the shuttle and to permit more ambitious tests and construction work there will be need for a space construction platform. It will be a large orbiting facility operated from a shuttle docked to one side of it. It will serve to greatly increase the ease of building truly large structures.

SIMULATE SPS SOLAR ARRAY PRODUCTION

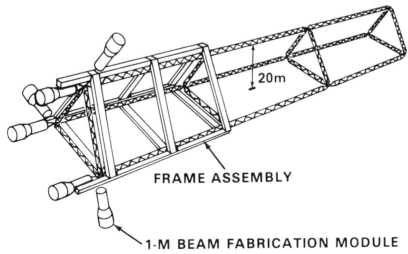

20m

FRAME ASSEMBLY

1-M BEAM FABRICATION MODULE

A twenty-meter-deep beam to be used in building the Solar Power Satellite (SPS) will be assembled from standard one-meter beams produced by beam-builders. (Courtesy NASA)

CONDUCTOR INSTALLATION

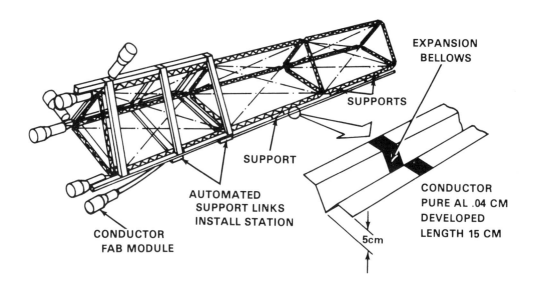

OBJECTIVE: — SIMULATE SPS POWER DISTRIBUTION SYSTEM INSTALLATION

TEST: — CURRENT DENSITY & HI VOLTAGE CHARACTERISTICS OF FABRICATED SYSTEM

As such large beams grow in length, other powersat systems will be added. An important problem is the installation of conductors, which will carry the electric current to the powersat transmitting antenna. Such conductors can be installed while the beams are manufactured. (Courtesy NASA)

As proposed by Tom Hagler and his associates in a study at Grumman in 1976, the platform will feature a 360-foot-long boom, which can rotate to any angle and will be outfitted with tracks along which manipulator arms and equipment carriers can travel. The boom will also carry a small cab resembling the cab of the familiar cherry-picker used by telephone linemen, with room for one or more crew members. The boom will serve in building the platform, which will be 236 feet long by 105 feet wide, in orbit 210 miles up. A large opening in the platform structure, 105 feet long by 79 feet wide, will serve for developing and demonstrating procedures for mounting solar arrays, thin-film mirror surfaces, wire mesh, and other components that will span wide areas of finished large satellites. To provide power, thirteen solar-cell arrays of the type used in the Solar Electric Propulsion Stage (SEPS) project will produce 250 kilowatts. Total platform weight will be 42 tons.

Three flights of the shuttle will serve to construct the platform. The first flight will deploy as a single unit a core module, which includes the control equipment. To this core module will be added a short section of the rotating boom, as well as one section of the solar array. The second flight will construct the inner area of the platform (105 feet square), the remainder of the long boom, and the rest of the solar array. The third flight will complete the platform structure and install the power distribution system, which will provide electricity not only for construction work, but for the shuttle itself. By relieving loads on the shuttle's own power system, the platform will greatly increase the time that a crew can stay on-orbit.

The resulting platform will provide a powerful and flexible means for assembling large communications satellites and other space systems. The first such system, to be constructed using the platform, may be a 330-foot-diameter radiometer, or microwave antenna, used to measure the natural radio waves generated by Earth's surface. It will detect the content of moisture in soil as an aid to crop forecasting. To build it, crew members working from the cab on the platform boom will install

SOLAR BLANKET INSTALLATION

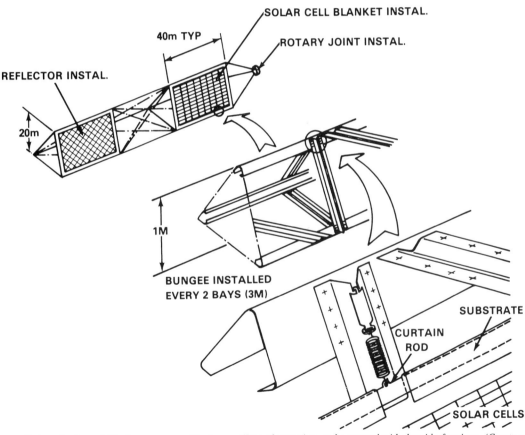

Installation of solar-cell blankets may resemble the unrolling of carpeting, to be secured with the aid of springs. (Courtesy NASA)

structural members as well as the reflector mesh surface. A crew of seven could build the radiometer in eight weeks.

Another early candidate for assembly at the platform would be a multibeam communications antenna of two-hundred foot diameter. It would provide 256 fixed beams aimed at specific ground stations, and 16 movable or scanning beams. The antenna would be built from 226 hexagonal elements, fitted together like floor tiles. Again, crews working from the boom cab would assemble these elements.

The most exciting uses of the platform, of course, would involve experiments on the construction of power satellites. Like the other space structures, powersats would be built of beams. But whereas ordinary beams would be one meter deep, fabricated with standard beam-builders, the beams for powersats would be twenty meters deep. Rather than being built with edge members and cross-braces cut and formed from sheet metal, the twenty-meter beams would be formed from assembled sections of one-meter beam. They would thus resemble the massive structural members of great suspension bridges that are built from lengths of standard steel beams, which alone suffice to build an ordinary highway bridge.

ROTARY JOINT INSTALLATION

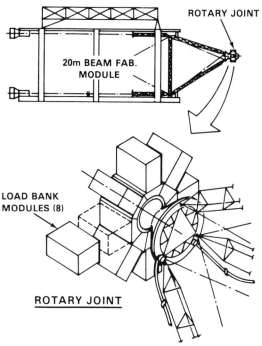

20m BEAM FAB. MODULE

ROTARY JOINT

LOAD BANK MODULES (8)

ROTARY JOINT

OBJECTIVE —

EVALUATE THE OPERATION OF SCALED DOWN ROTARY JOINT MECHANICAL & ELECTRICAL INTERFACES.

TEST REQUIREMENTS —

- VERIFY INSTALLATION PROCEDURES
- VERIFY MECHANICAL OPERATION OF THE BALL JOINT DRIVE AND SUSPENSION SYSTEM.
- VERIFY POWER TRANSFER THROUGH THE ROTARY JOINT 40KVDC & 7.75 AMPS/CM^2 PER BRUSH

An important test will involve the rotary joint, which in a powersat will support the transmitting antenna while allowing the passage of massive electric currents. (Courtesy NASA)

The OCDA's long boom can serve to test a powersat's microwave beam system. Here the boom has been fitted with klystrons (microwave generators), while a test satellite hovers some distance away. As the boom sweeps from side to side, the satellite will be used to study the characteristics of the transmitted microwave beam. (Courtesy NASA)

To build sections of such great beams, the platform would mount a construction frame carrying six standard one-meter beam-builders. The first test specimen would be 246 meters long and, when completed, would be suspended beneath the platform, then shaken for vibration tests.

The platform will support other tests of critical questions in powersat construction. A key problem area involves installation of electrical conductors. In a full-size powersat such conductors will stretch for miles, carrying high voltage and current. It will be possible to test means for fabricating and installing such power conductors simultaneously with the construction of their supporting twenty-meter beams.

Another problem area involves installing large panels or blankets of solar cells. A promising approach is to produce such solar panels as roll-up arrays, attached at their edges to a crossbeam of a twenty-meter main beam. As the main beam extends in length during construction, the solar array unrolls and can be fastened at the sides. Still another area for study involves formation of tightly

The Space Spider can produce structures by laying down aluminum along a path formed on its previous pass. (Courtesy Marshall Space Flight Center)

focused microwave beams for power transmission to Earth. To check this out, the platform's boom can mount an array of klystrons or microwave generators. A small satellite will operate with a beam-mapping system at a distance of several tens of miles. As the boom swings from side to side, the satellite will study the characteristics of the transmitted microwave beam.

Moreover, while beams (both of microwaves and of aluminum) are the key to construction of power satellites, there are other ways to build large space structures. When the structure has a simple shape, as many will, there is the opportunity to use the Space Spider, invented by J. D. Johnston of NASA-Marshall. This novel construction technique lays a braced aluminum structure built from fifteen-foot-diameter rolls of aluminum, around the path created on its last pass. The simplest such structures are spiral-shaped and have the pattern of a lawn mower cutting a spiral path to cover an entire back yard. The Space Spider can advance at twenty feet per minute, thus constructing an eight-hundred-foot diameter structure in only seventy-two hours.

In the more distant future, the Space Spider may be the best way to build a true space colony. The best-studied colony design, the Stanford torus, features a living area shaped like a bicycle inner tube, 400 feet wide by a mile in diameter, rotating within a ''tire'' 6 feet thick and filled with lunar sand, as a radiation shield. Six spokes, 50 feet in diameter, lead from the colony rim to the hub: a 425-foot-diameter sphere. All these structural shapes must be strong and airtight, like the hull of an airliner; and none lend themselves in a simple way to being built from light, open latticeworks of beams. The Space Spider may fabricate their forms directly, in an elegant design resembling the preformed fiberglass hulls of catamarans.

Opposite: The explosion of a supernova, while incinerating nearby planets, also injects th heavy elements into interstellar space from which new planets will someday form. *(Courtes Don Dixon)*

Out of primordial clouds of gas and dust, stars and planets form. *(Courtesy Don Dixon)*

The dream of those who would colonize space: a world set free. Foreground, astronauts approach one of several powersats which can be seen. Below, Earth's night side blazes with city lights, whose energy comes from powersats. Agricultural land encroaches into the Sahara. On the Moon, a dot of light marks the lunar base. *(Courtesy Don Dixon)*

Above: A space freighter docks at the central receiving station of the power satellite construction center. *(Courtesy Boeing Aerospace Co.)*

Opposite top: Single-stage-to-orbit craft ascends. *(Courtesy Don Davis)*

Opposite bottom: Sports in zero-g. *(Courtesy Ames Research Center)*

Above: "Sunrise on Akir." *(Art by Don Dixon, courtesy New World Pictures,© 1979)* Rendered as a design concept for the movie *Battle Beyond the Stars.*

Opposite top: "Roche's Limit." *(Courtesy Don Dixon)*
Opposite bottom: "Antares Rising." *(Courtesy Don Dixon)*

Use of Space Spiders to build a space colony of the Stanford torus type. In the foreground mobile teleoperators carry rolls of aluminum to restock the Spiders' supplies. Detail shows a Spider laying down the hull of the colony, which has the shape of a bicycle tire. Central disk structure will carry solar arrays. (Courtesy Marshall Space Flight Center)

Such projects may be far away, but beam-builders and the shuttle-supported construction platform lie just around the corner (or the next presidential election, if you will). They will take us to the very verge of being able to build power satellites. They will not take us all the way, for the shuttle will never serve the immense logistical needs of a true powersat. But once these systems have been built and proven, once their experiments have succeeded and the world has grown accustomed to the communications platforms and other large structures they will build—then the needed confidence will be in hand for the next step.

That next step will be to go forward and build the vast orbiting station needed for building powersats. As long as space construction is constrained by the limitations of the shuttle, it will have the character of a machine shop in someone's garage: Its equipment will call for only a few people, for only occasional use. In preparing for the powersat, the need will be for a space factory. It will have many items of equipment, at diverse work stations, most of which will see nearly constant use. The number of people in space will leap from a few to several hundred; their stay-times will go from weeks to months. For all this, the jobs and activities in this factory will first be tested in the shuttle era, in the orbiting construction base.

129

posite: Starship arriving at a distant star system. *(Courtesy Don Dixon)*

CHAPTER **7**

Robots and Other Space Workers

In the science-fiction movies of recent years, among the most appealing characters has been the robot Artoo-Deetoo (R2D2) of *Star Wars*. It is hard not to refer to such a robot as "he," and in his diminutive way (he looked rather like an overgrown canister-type vacuum cleaner), he performed prodigies of exertion in the service of the beautiful Princess Leia, leader of the revolt against the galactic overlord Darth Vader. But there was one thing wrong. *Star Wars* was set amid an interstellar civilization technically in advance of our own; but to anyone conversant with modern computers, Artoo-Deetoo would be lamentably out of date.

For Artoo-Deetoo could not speak. He communicated by means of chirps and whistles, * a fact that immediately dates him. Computer speech is already a reality, and what is more, has been so for some time. Indeed, it is likely that by now most Americans have had the experience of hearing a computer talk, even if few have been aware that that is what they have heard.

Pick up the phone and dial a number that is out of service. In many cities, doing this will produce a message: "The number you have reached, 123-4567, is not in service. Please check the number and dial again." The message will sound like a recording, but there's a catch; it specifies the number dialed. The phone company can't keep thousands of separate recordings of out-of-service numbers, not to mention keeping them up to date. The answer is that the message is not a recording. It is a computer synthesis of human speech.

How can a computer synthesize human speech? The technique is actually no more marvelous than the means whereby, via ink marks on paper, literate people can learn of things they never

---

130   *In fairness, his robot friend See-Threepio (C3PO) could not only speak but could, the occasion demanding, wax eloquent.

experienced. The "alphabet" of computer speech is straightforward. It is no more than the sounds, or phonemes, which make up human speech. There are about forty-five of them: long and short vowels, hard and soft consonants, and the like; they are listed in any dictionary. In ordinary speech or conversation, one hears or produces ten to twenty such phonemes per second.

Each of these phonemes can be synthesized from a combination of only a few standard sounds: tones of different frequencies, which can be added or combined by means of electronic filters, together with hissing sounds. A hissing or sibilant sound, known as white noise, is essential for phonemes like *s, t, th, b,* and others. Then, by means of a method of computer programming known as LPC (Linear Predictive Coding), one can very efficiently specify the way the electronic filter is to combine the various tones and the sibilance, while allowing for changes of tone, so that computer speech can sound like a man or a woman, with natural-sounding rises and falls of inflection. A flat-sounding monotone that grates on the ears is thus avoided.

To most adults, the idea of hearing a computer speak may seem bizarre, even frightening. But the next generation of children may well grow up regarding it as part of their everyday worlds. Texas Instruments, a leading manufacturer of advanced electronics, recently announced a new voice synthesizer which uses only three integrated-circuit chips, giving it roughly the complexity of the familiar pocket calculator. One of its uses is in a new $50 childrens' game, Speak and Spell.

Speak and Spell has a keyboard for letters and numbers, a calculatorlike display—and the synthesizer. For instance, it will say to a child: "Spell CAT." The child responds by pushing buttons on the keyboard. There is a pause. Then, if the child is right, the box says, "You are right," and goes on to the next word. If the answer is wrong, the box says, "No, that's incorrect." Then the display flashes the right spelling, and the synthesizer spells it out verbally as well.

If computers can talk, can they also listen? Experiments in teaching computers to recognize human speech go back at least twenty years but have had mixed success to date. The problem is one of having a computer match a pattern of frequencies and overtones, representing a word spoken to it, with one of a large number of patterns stored in its memory. If it is to understand ordinary speech, it must be able to match such patterns very rapidly, searching over many possible matches each second. With the fast computing times and large memories of modern computers, this can indeed be done. Bell Labs has a machine today that will recognize spoken words with up to 98 percent accuracy. To do this, a person first speaks each of several thousand words, a dozen or so times each, and the computer records what it hears in its memory. The subsequent word-recognition is keyed to the speech of that particular individual, so that it cannot understand the speech of a variety of different people.

This is in any case rather a brute-force procedure, since it ignores such clues as syntax and grammar. Still, it can go further than this. By the early 1980s, as memories improve, computers will be storing patterns for up to half a million words to be recognized when spoken. Each word will be stored as twenty or thirty general patterns or variations for a vocabulary larger than most people possess. Because the patterns will not be keyed to any one person, the machine should be able to do a creditable job of understanding anyone's ordinary speech. In any case, if the machine fails to catch a word or phrase, it can do what anyone else would do in that situation. It will simply say, "Excuse me, I didn't catch that."

With these abilities computers will be well along in their use of artificial intelligence, or AI. AI has intrigued computer scientists for quite a few years. As early as 1965 an AI program allowed a computer to play championship-level checkers. More recently chess-playing computers have come into their own. For $300 any chess buff today can buy BORIS, which offers play at several levels of competence. It can do things that few human opponents would accept; thus, when BORIS backs you into a corner, just change places with him and see how he battles his way out of his own trap. While BORIS is thinking, he electronically moves pieces around the board, analyzing the effect of each move he considers. He will anticipate opponents' moves, displaying such comments as ''I expected that'' or ''Is this a trap?''

Chess is an excellent problem for AI since there are many more possible games (and moves) than there are electrons in the Universe. Hence a chess-playing computer must be able to recognize patterns and eliminate huge numbers of inappropriate possibilities. This attribute, of course, is what is needed in order to recognize ordinary human speech. To be sure, computers can do rather more than play chess. As early as a decade ago, Joseph Weitzenbaum of MIT devised a program called ELIZA, with which one could converse with a computer in English, using a keyboard to type in statements. He also gave the computer another program, DOCTOR, designed to permit it to play the role of a psychotherapist. In his book, *Computer Power and Human Reason,* he records the following exchange:

Men are all alike.
IN WHAT WAY
They're always bugging us about something or other.
CAN YOU THINK OF A SPECIFIC EXAMPLE
Well, my boyfriend made me come here.
YOUR BOYFRIEND MADE YOU COME HERE
He says I'm depressed much of the time.
I AM SORRY TO HEAR YOU ARE DEPRESSED
It's true. I am unhappy.
DO YOU THINK THAT COMING HERE WILL HELP YOU NOT TO BE UNHAPPY
I need some help, that much seems certain.
WHAT WOULD IT MEAN TO YOU IF YOU GOT SOME HELP
Perhaps I could learn to get along with my mother.
TELL ME MORE ABOUT YOUR FAMILY
My mother takes care of me.
WHO ELSE IN YOUR FAMILY TAKES CARE OF YOU
My father.
YOUR FATHER
You are like my father in some ways.
WHAT RESEMBLANCE DO YOU SEE
You are not very aggressive but I don't think you want me to notice that.
WHAT MAKES YOU THINK THAT I AM NOT VERY AGGRESSIVE
You don't argue with me.
WHY DO YOU THINK I DON'T ARGUE WITH YOU
You are afraid of me.
DOES IT PLEASE YOU TO BELIEVE I AM AFRAID OF YOU
My father is afraid of everybody.
WHAT ELSE COMES TO MIND WHEN YOU THINK OF YOUR FATHER
Bullies.
DOES THAT HAVE ANYTHING TO DO WITH THE FACT THAT YOUR BOYFRIEND MADE

YOU COME HERE

When a computer does psychoanalysis, or when it plays a good game of chess, there is obviously the question: Can it match the thinking ability of a human being? The important feature of human intelligence is not that ordinary people can match a computer when it comes to number-crunching. The mathematician Karl Friedrich Gauss committed logarithm tables to memory, the better to aid his studies, and there have been occasional calculating prodigies who could multiply twenty-digit numbers in their heads. But people usually outshine a computer at dealing with patterns and symbols, particularly when the patterns are complex. Why? Because the amount of memory available to an ordinary human is something like a trillion bits, or twenty thousand books the size of the Bible. (The King James Version has 773,693 words.)

Until recently there was no prospect that a computer could match this. True, such quantities of memory could be stored on magnetic tapes; but it would be a slow, inefficient process to mount tapes and have the computer search them through. Also, pattern-recognition programs have not been powerful enough to allow computers to make effective use of such reserves of memory. But the video disk has brought new hope. To electronics buffs, the video disk is already familiar; it is about the size of an ordinary long-playing record but records and plays not music but rather TV shows. A single such disk can store up to four hours of TV viewing. Used as a means of storing data, such a video disk can readily store a trillion bits. With appropriate coding and organization, any part of this information can be made available to a computer in less than a tenth of a second.

There is little doubt that AI programs will grow smarter, more subtle. Already MIT has a program whereby a computer can learn about its world as would a child. This concept-learning program can infer the idea of, say, an arch simply by being shown a set of pictures of arches and non-arches. A common-sense-reasoning program written at Stanford Research Institute learned to extend its initial solution to a particular problem to a more general solution applicable to a range of problems.

One trend in AI will be to develop ''knowledge about knowledge''; that is, models of what a computer knows, so that it can select or anticipate an appropriate response. To a degree, BORIS does this in chess when it predicts correctly an opponent's move and flashes, ''I expected that.'' In programs such as ELIZA, which deal with written or spoken English, a knowledge of grammar and syntax will obviously permit advances over the brute-force recognition or matching-up of word patterns. Even more subtlety will be available through the programming of rule-making rules. Thus, a computer may start with a set of basic rules for understanding conversation or English text. However, with these rules as examples, it will be able to formulate new rules and to try different sets till it gets the best response and can interpret its material most fluently.

Within a decade all these trends may culminate in a computer that can read the texts of thousands of books and *interpret* them. It will be able to talk and to listen. One could thus converse with it as with the brightest and most well-read of companions. It will be the Speak and Spell game, but at an unimaginably richer and more complex level.

By the time AI reaches this level, it will be possible to anticipate a day when robots and computers will fulfill all the functions and perform all activities for which we now believe humans will be needed in space. Such computers will even check out one another's performance, carry out maintenance, or make repairs, removing faulty electronics units and replacing them with good ones. This will actually be a continuation of trends of past decades.

In the 1950s it would scarcely have been believed that the exploration of Mars would fall to such craft as Viking. Even in the 1970s there was much pointing with pride at the lunar explorations of   *133* astronauts, who could select geological samples of interest. Today this is ancient history. When anticipating similar explorations of Mars, it is generally agreed that there will be mobile robot vehicles resembling Viking spacecraft with tank treads. Ranging widely over the Martian surface for months or years, they will stop frequently to make measurements and chemical analyses. The most interesting rock samples will be loaded aboard rockets and returned to Earth under automatic control. From a control center in Pasadena, the world's best geologists will learn more about Mars than the Apollo teams learned by actually visiting the Moon.

As in exploring Mars, so one day will sophisticated robots assemble entire power satellites; but that day still appears far off. If the power satellite becomes a major project before century's end, almost certainly the automated systems used will be far less sophisticated. Like power shovels and other road-building equipment, there will be the beam-builders and cranes or manipulators described in Chapter 6, each requiring control by an operator. Other space workers will operate the boom-supported cherry pickers or remote work stations. The highest degree of sophistication will be in the remotely controlled robots known as teleoperators.

The Space Spider will be one such teleoperator. Under control of an internal computer, it will lay down metal structures by following along the periphery of a shape already built. It will do this until it is commanded to stop or its rolls of sheet metal run out. A more interesting type of robot will be the free-flying teleoperator. It will be equipped with manipulator arms and "end effectors"—robot

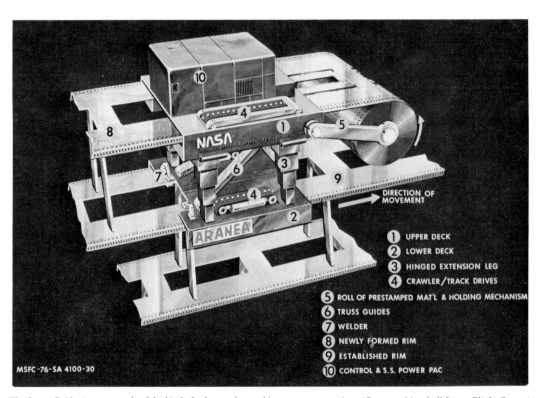

DIRECTION OF MOVEMENT

1. UPPER DECK
2. LOWER DECK
3. HINGED EXTENSION LEG
4. CRAWLER/TRACK DRIVES
5. ROLL OF PRESTAMPED MAT'L & HOLDING MECHANISM
6. TRUSS GUIDES
7. WELDER
8. NEWLY FORMED RIM
9. ESTABLISHED RIM
10. CONTROL & S.S. POWER PAC

MSFC-76-SA 4100-30

134   *The Space Spider is an example of the kind of robots to be used in space construction. (Courtesy Marshall Space Flight Center)*

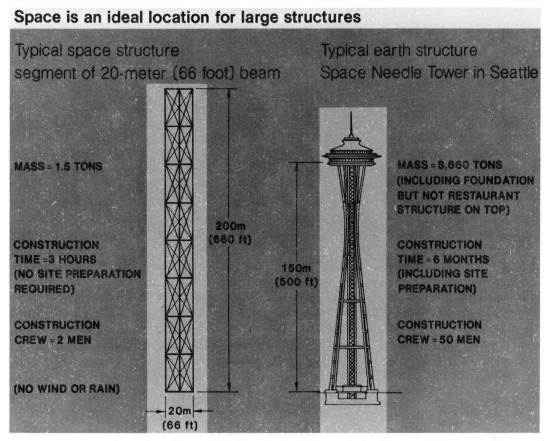

Space is an ideal location for large structures

Typical space structure
segment of 20-meter (66 foot) beam

Typical earth structure
Space Needle Tower in Seattle

MASS = 1.5 TONS

CONSTRUCTION
TIME = 3 HOURS
(NO SITE PREPARATION
REQUIRED)

CONSTRUCTION
CREW = 2 MEN

(NO WIND OR RAIN)

200m
(660 ft)

150m
(500 ft)

20m
(66 ft)

MASS = 8,660 TONS
(INCLUDING FOUNDATION
BUT NOT RESTAURANT
STRUCTURE ON TOP)

CONSTRUCTION
TIME = 6 MONTHS
(INCLUDING SITE
PREPARATION)

CONSTRUCTION
CREW = 50 MEN

Comparison of space and Earth construction. (Courtesy Boeing Aerospace Co.)

hands, if you will. It will also have TV cameras and lights and a rocket motor with propellant. Its operator will use its TV cameras to see what it sees, and by remote control the operator will steer it to a particular work station. There the robot will be put to work using tools or other equipment it may carry, while the operator sits in the comfort of his control center, projecting his skills through a radio link.

Once there is good understanding of the construction equipment to be used, it is a straightforward matter to determine how large a work force must be supported in space in order to build power satellites. Even without robots of extraordinary sophistication, even with beam-builders and similar items that call for the attention of skilled human operators, the construction rates can be astonishing. This is because of the ease with which large structures can flow from a source, being very light in design and requiring no elaborate bracing.

The Space Needle in Seattle, a relic of the 1962 Worlds Fair, is a good example of the difference between space and earth construction. Leaving aside the revolving restaurant on top, it is 150 meters high. It has a massive concrete foundation, and with the foundation but without the restaurant, it weighs 8,660 tons. It took six months to build with a construction crew of fifty. By contrast, a similar-size space structure built of beams would be 200 meters long by 20 wide. Such a structure, indeed, would be a section of twenty-meter beam, described in Chapter 6 as the main structural element of a powersat. It would weigh 1.5 tons and take three hours to build using a crew of two.

So the work force will not be large, by standards of even a modest earthside construction project. Still, it will dwarf all previous notions of the size of space crews. It will be, indeed, the first large-scale use of people in space. There will be about five hundred workers in a variety of categories. At Boeing, where Eldon Davis and Keith Miller have studied the problems of power satellite construction, the space jobs to be filled are believed to include

Managers	Communications equipment operators
Supervisors	Traffic controllers
Clerical staff	Space transportation vehicle maintenance and
Planners	operations technicians
Beam machine operators	Hotel keepers
Crane/manipulator operators	Utility operators
Solar array deployment machine operators	Food service personnel
Antenna subarray deployment machine	Physicians
operators	Dentists
Maintenance technicians	Paramedics
Cargo handling equipment operators	Chaplains
Test and quality control specialists	

These jobs are similar to those in present-day industry. The work of a crane/manipulator operator would resemble that of manipulator operators in the nuclear industry, where radioactive fuel rods must be handled remotely, behind thick shielding. The operators for automated equipment like the beam machines and deployment machines will have work similar to that of operators of automated wing riveter machines in the aerospace industry or of paper-making or bottle-making machines. The people who do maintenance, cargo handling, communications, and traffic control will likely have done very nearly the same things at airports or air-freight terminals. The management, supervisory, and staff jobs will be like those found in any earthside manufacturing plant.

When these jobs open up, applicants will not have to be superior people, like astronauts. They will be chosen for being adventurous, ambitious, hard-working, intelligent, and having a strong commitment to excellence in their work. Requirements on their physical condition will be minimal, most notably that they be in good health, capable of tolerating several g's of acceleration, minimally susceptible to motion sickness, and not too far from the general population in height and weight. There will be psychological tests, like those used to select crews of submarines, in order to detect applicants who could not stand the cramped quarters and unfamiliar environments of space.

Equal opportunity laws will be used in hiring; there will be both men and women in the space workers' employment line. There will probably be at least two complete crews, each numbering some five hundred: one to work in space and the other to be on Earth between trips to space. If powersat construction rates are stepped up, there will be more crews as well, and there will probably be tens of thousands of applicants for these prized jobs.

The training of selectees will start soon after they are hired and will call for an enormous investment in facilities and simulators. As in any complex project, the size of the training crews will be at least as large as that of the crews they are training. There will be classroom studies and plenty of

Space construction workers will train in simulated weightlessness in the Neutral Buoyancy Facility. (Courtesy Marshall Space Flight Center)

practice sessions with the equipment they will use, ranging from the most advanced teleoperators to the stew pots used by the cooks. The most realistic forms of training, and therefore the most important, will involve simulating actual on-the-job work activities and weightlessness.

For the on-the-job simulations, the equipment-operator crews and their supervisors will have control cabs, cherry pickers, and instrument panels just as they would have in space. Within the crew members' fields of view will be, not actual space hardware under construction, but rather full-color TV screens, seven feet across or even larger. These will be linked to a powerful computer capable of generating TV images of work in progress. The computer will produce a continuously updated **137** representation of the construction and will generate the TV displays that will show each worker what in fact would be seen if actual construction were taking place.

The actions of individual workers will not produce real beams or solar-cell arrays. Instead, they will generate signals to the computer, which will interpret them as actions that would have produced so many meters of beam or array. So far as concerns actual production, the trainees will be like the tailors who stitched the Emperor's New Clothes in the Hans Christian Andersen tale, whose needles held no thread, whose looms bore no draperies. But when a beam-machine operator switches his controls, he will have the satisfaction of seeing lengths of beam energe on his TV screen. If someone else is to attach the edge of a solar-cell array to the length of the beam, there it will be on the TV screen, as expected. When production equipment would run low on supplies because of long use, teleoperator controllers will simulate the reloading of magazines and stores.

This type of large-scale simulation will not only be a most effective way of training; it will help assure program managers that the powersats will in fact be built according to schedule. For weeks or months on end, crews of trainees can follow their work schedules, as major sections of power satellite emerge and take shape entirely within the computer. Someone may fall sick and be absent; a replacement will have to carry on. Groups of people will be rotated off the job, as if to return to Earth; the new crew members will pick up where they left off. Equipment can be made to shut down or malfunction; managers and foremen will have to devise means of working around the problems to keep production on schedule. People will grow bored, fatigued, irritable; they will demand coffee breaks or times when they can relax and stretch their legs. Inevitably, some otherwise highly qualified people will find the job of a space worker is not for them. It will be much easier to let them go when this involves walking out a door into the Texas sunshine, rather than being returned from earth orbit.

Most of these simulations will be in normal gravity, but training for weightlessness will be very important, especially for maintenance people and for those whose work will require great skill and care, in contrast to the more routine activities of the equipment operators. Training for weightlessness, of course, dates back to the 1950s. For at least that long there have been jet aircraft flying special maneuvers to give their passengers thirty seconds or so of the real thing. Since the mid-1960s, would-be astronauts have trained in a much more convenient and better way: under water.

At NASA's Marshall Space Flight Center is a large swimming pool, seventy-five feet in diameter and forty feet deep, containing over a million gallons of water. This is the Neutral Buoyancy Facility. Properly weighted, an astronaut under water will float freely within it, as if in zero-g. What's more, the tools and equipment he handles can also be made buoyant. Aluminum beams can have small blocks of styrofoam attached at their ends; the beams then behave very much as if in space. These underwater exercises can go on for three hours or longer.

The large size of the tank allows people to work with full-scale structures, just as there would be in orbit. Most astronauts training in it have worn complete space suits, which double as diving suits when weighted with lead. However, the upcoming Spacelab flights will have astronauts working in shirtsleeves, and underwater training without pressure suits has been introduced as part of this program. In these "shirtsleeve simulations," people use only ordinary scuba equipment with a face mask but no flippers. It would be considered bad form to actually swim under water; instead, people move about by grasping hand-holds or climbing along the ladderlike beams of space structures.

The resulting training is indeed realistic. The Neutral Buoyancy Facility saw extensive use during the first Skylab flight in 1973. There the crew leaned the procedures that enabled them to save their damaged space station. When *Aviation Week* Editor Craig Covault took part in underwater training, he

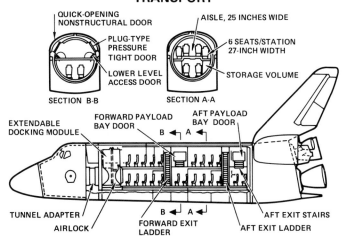

INBOARD PROFILE; 74-PASSENGER ORBITER TRANSPORT

QUICK-OPENING
NONSTRUCTURAL DOOR

AISLE, 25 INCHES WIDE

PLUG-TYPE
PRESSURE
TIGHT DOOR

6 SEATS/STATION
27-INCH WIDTH

LOWER LEVEL
ACCESS DOOR

STORAGE VOLUME

SECTION B-B

SECTION A-A

EXTENDABLE
DOCKING MODULE

FORWARD PAYLOAD
BAY DOOR

AFT PAYLOAD
BAY DOOR

TUNNEL ADAPTER

AFT EXIT STAIRS

AIRLOCK

FORWARD EXIT
LADDER

AFT EXIT LADDER

Space workers may fly to orbit aboard a rocket such as the Shuttle-Derived Vehicle. The cargo bay of the shuttle orbiter could be equipped to carry seventy-four passengers. (Courtesy Rockwell International Corp.)

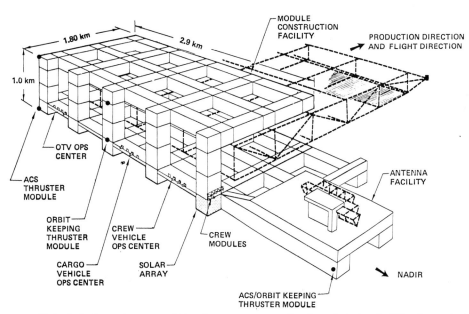

1.80 km

2.9 km

MODULE
CONSTRUCTION
FACILITY

PRODUCTION DIRECTION
AND FLIGHT DIRECTION

1.0 km

OTV OPS
CENTER

ANTENNA
FACILITY

ACS
THRUSTER
MODULE

ORBIT-
KEEPING
THRUSTER
MODULE

CREW
VEHICLE
OPS CENTER

CREW
MODULES

CARGO
VEHICLE
OPS CENTER

SOLAR
ARRAY

NADIR

ACS/ORBIT KEEPING
THRUSTER MODULE

Powersats would be built in an immense orbiting factory or construction base featuring facilities for producing both transmitting antennas and solar-blanket modules. The construction base would house 500 workers and build one powersat per year. Abbreviations: OTV, orbital transfer vehicle; OPS, operations; ACS, attitude control system. Note the relatively small solar array which provides power for the center. (Courtesy Boeing Aerospace Co.)

Traffic at the construction base. At left a space freighter unloads its cargo; such freighters arrive about once a day. At right a shuttle orbiter prepares for return to Earth with passengers. Much of the base consists of a delicate tracery of lightweight beams. (Courtesy Boeing Aerospace Co.)

found few clues other than bubbles from other divers that the activities were indeed being conducted under water and not in space.

In time all training exercises will come to an end, and the space workers will be sent to begin the jobs for which they were hired. To get to space, there will be what in the era of the power satellite will count as a golden oldie—the space shuttle. With an advanced booster giving lower cost, the shuttle will be just the right size to carry seventy-five to one hundred crew members to and from orbit. For this, the shuttle's payload bay will hold a passenger module, somewhat resembling a section of tourist-class seating from an old-style airliner like the 707 or DC-8. The seating will likely be five abreast, and there may be no windows, stewardesses, or choice of hot dinners. The passenger module will be like a sealed and enclosed Greyhound bus, loaded by crane aboard an aircraft to be delivered as air freight. The passengers' first experience with space flight thus would present few amenities beyond what would be found in a crowded subway stalled underground, but mercifully the flight will be brief. And there will at least be the opportunity to install a TV screen to show views of the world outside and of the construction base as the spaceliner approaches.

The construction base will be an immense open structure of beams extending for miles. Here and there the discerning eye will pick out the rocket transport centers, perhaps with spacecraft moored beside them. There will be the centers for receiving the great cargo rockets from Earth, for unloading and storing their massive payloads, or for serving as terminals for passenger traffic. The power plants for the construction base with their purple solar arrays will be visible, as will the clusters of modules used for crew quarters. But only close in, when the base appears as a fathomless immensity of beams, will anyone recognize the tiny and widely dispersed work stations scattered amid the vastness.

At the construction base the people will live in crew modules somewhat like the Skylab space station. These will have been built and fitted out on the ground and carried to orbit by means of the large cargo craft. They will be larger than Skylab—say, fifty-five feet in diameter rather than twenty-two, with seven levels with ceilings at little more than seven feet. But whereas Skylab held only three people, each of these units will hold something like a hundred. Three levels will be personal quarters. Two others will serve for storage and for the heating, cooling, oxygen, and waste-control systems. An entire level will be fitted out for the galley and the cafeteria or dining hall, and the seventh level will serve as a zero-g gymnasium.

How will these people live? This question distinguishes the power satellite enterprise from science fiction for science fiction writers rarely have to worry about how their heroes get their laundry done or wash their dishes. The experience of Skylab and the space shuttle will be valuable here. Skylab carried a fold-up shower system, which its astronauts much appreciated. The shuttle carries a sit-down toilet, which can be used in space by both men and women in privacy. In both these items, currents of air mix with sprays of water as a substitute for gravity. There will be a water-reclamation system to recycle the waste water from showers and toilets, as well as from sinks and kitchen galleys, since this water would not be for drinking.

The senior managers will have private suites, since rank hath its privileges even in space; but most of the people will sleep dormitory-style. They will zip themselves into "sleep restraints," comfortably padded, loose-fitting affairs mounted to the bulkheads. It will be possible to provide privacy by means of curtains, and there will be ample drawer or cabinet space for the clothes and belongings of each crew member. When the people get up in the morning, the resemblance to a college dorm will be evident: people lining up to use the shower or to brush their teeth at a sink, or rubbing the sleep from their eyes as they prepare to shave. (The electric razors will come with tiny vacuum cleaners to vacuum up the bits of hair.) Morning coffee will be served in the dining area.

The care and feeding of the crews will receive careful attention, since appropriate concern for the inner man will be one of the ways to maintain morale. Certainly a hot, appetizing, home-cooked meal will be welcomed by tired construction workers at the end of a shift, just as on Earth. The food will include plenty of fresh water, vegetables, meats, milk and eggs, and much else, all delivered fresh from Earth. The frequent cargo flights, one or two per day, will carry five hundred tons of cargo each, so that fresh food will be a minor item on the cargo manifests. Indeed, since the value of the workers' labor will be some thousands of dollars per hour, it will be no great cost to have frequent servings of lobster, fine steaks, or even French or Chinese cuisine. For some people, the most lasting memory of construction days may be the filet mignons. There will, of course, be a full-time cafeteria staff to prepare the food and serve it. But for all that, even the best turkey dinner will not taste the same as it would on Earth. In zero-g there are shifts in body fluids which result in sinus and nasal congestion. Without a sense of smell, foods just don't taste the same.

In the dining areas, as well as in the recreation areas, one of the most crew-pleasing features will be plenty of large windows that face Earthward. The Skylab astronauts all acclaimed the large window

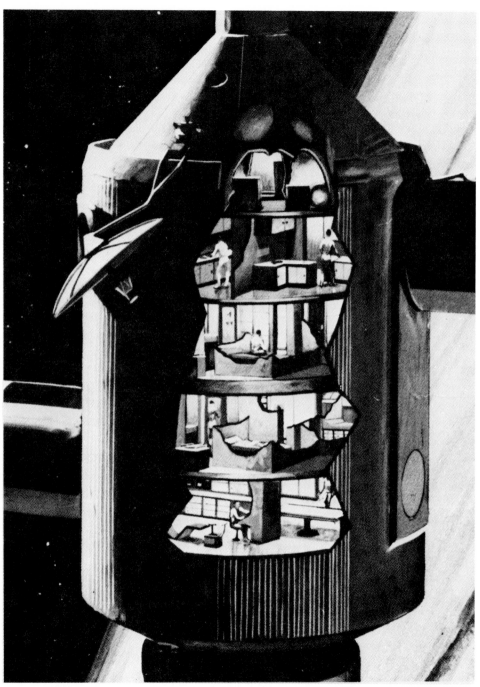

The construction people would be housed in 100-person "hotels," fifty-five feet in diameter, which would resemble this thirty-three-foot-diameter space station studied extensively in 1970. (Courtesy Rockwell International Corp.)

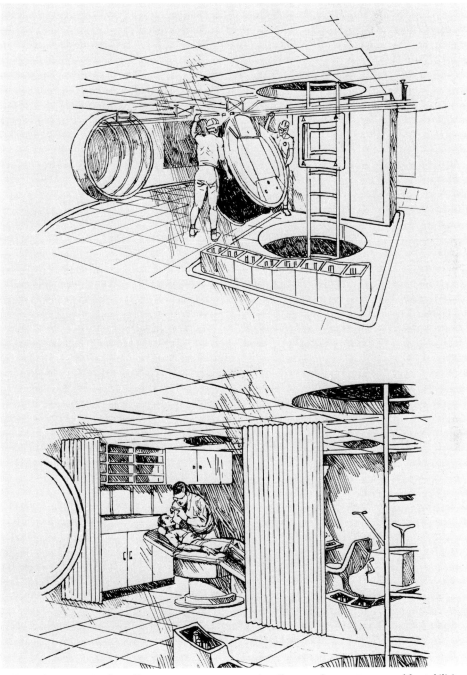

Everyday life in the construction base. Top, *two crewmen prepare to install a control-moment gyro, used for stabilizing motion.* Bottom, *a dental checkup. (Courtesy Rockwell International Corp.)*

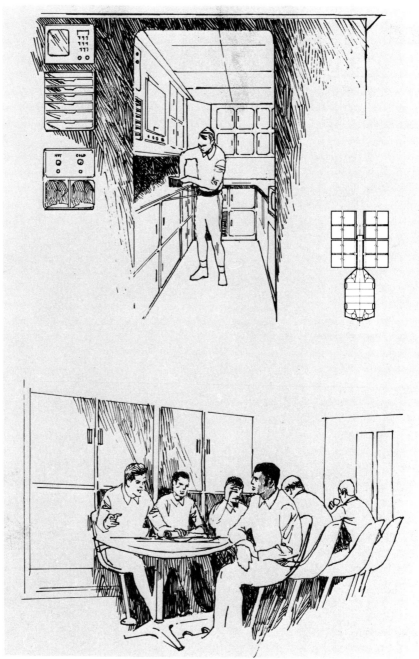

Culinary arts in the construction base. Top, preparing a late-night snack. Bottom, the fellowship of a wardroom at dinner time. Note the semicircular cutouts at the base of the table. Crewmen can stick their feet in them to remain in place despite their weightlessness. (Courtesy Rockwell International Corp.)

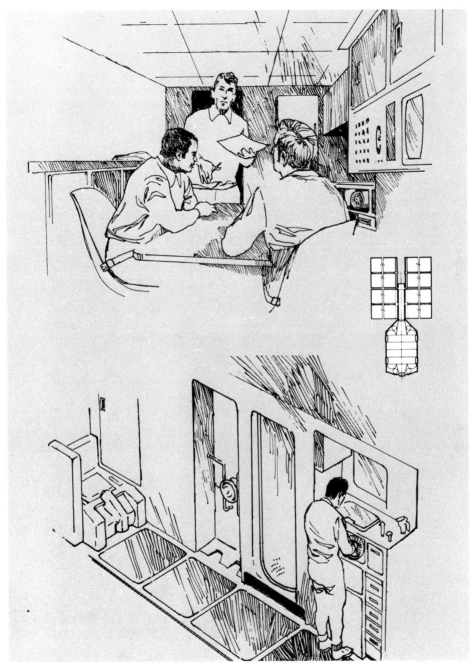

More daily life. Top, *a briefing to senior management.* Bottom, *a zero-g bathroom. Facilities include a toilet (at left), a suction urinal adjustable in height, a shower, and a sink. To use the sink, one sticks his hands through rubber sleeves; the faucets are on the inside. A window at the top lets people see what they are doing. (Courtesy Rockwell International Corp.)*

in their wardroom. For them, Earth-watching from the wardroom was one of their favorite pastimes. They all wished that the window had been bigger and that they had had more clearance as well as hand-holds, so that they might change their positions as Earth passed below. They recommended that there be an observation bubble.

There will be many other recreations besides Earth-watching. There will be daily movies, college extension courses, videotape TV, and saunas. The space crews will want libraries for books and music, telescopes, and regular mail and telephone service. In addition, since some couples will get married and others will want to act as though they were, it will be important to have a few padded rooms (possibly equipped with stereo and other amenities) where lovers can do what they feel is appropriate.

There will also be recreations of a more conventionally athletic nature. The periphery of the recreational area can readily be a jogging track; there was just such a track aboard Skylab. Interestingly, a jogger running at ten miles per hour would generate one-quarter g of artificial gravity. A sprinter at twice that speed would enjoy a welcome reacquaintance with normal gravity. There would be a large open area for such activities as zero-g acrobatics and gymnastics or handball. In addition, the people would welcome the kind of exercise equipment used aboard Skylab. This equipment included a treadmill for running, a stationary bicycle, and arm exercisers.

The five large crew modules will not be the only ones. The most senior people on the project may

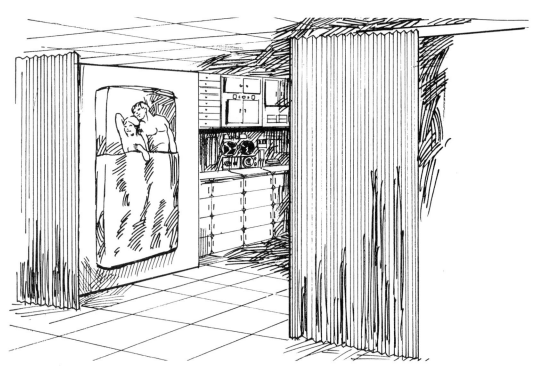

Sleeping facilities will resemble sleeping bags more than they will resemble beds as we know them, but the men and women working in space will nonetheless find them quite comfortable. (Courtesy Don Dixon)

"The Man and the Machine": The President of the United States at the controls of a manipulator used in powersat construction. With him are the First Lady and a technical specialist. The first presidential trip to space may be occasioned by the powersat effort. (Courtesy Don Dixon)

have a special module all their own, which may also have rooms for visiting VIPs. Just as the first presidential trip overseas came when Teddy Roosevelt visited the Panama Canal construction in 1906, so a century later the power satellite project may give the occasion for the first flight by the president to space.

Another module will be set aside to serve, at least in part, as the hospital and sick bay. It will provide as well for dental offices—and a morgue. There will be a training and simulation module to work with new construction equipment and techniques. Some of the new people who come to the project in midstream will receive training here. A maintenance and rest module will provide the opportunity to work on large items of construction equipment or spacecraft components in a normal atmosphere. The operations module will house the control center and will serve as headquarters for the managers and administrators. Perhaps this will also be where they will live and where the VIP suite will be located. Finally, a transient crew quarters module will provide for crews who are to be transported to and from their work locations.

There will be several different kinds of crew transportation. The space shuttle and its passenger modules will carry crews down as well as up. A few dozen picked crew members will staff a lonely outpost in geosynchronous orbit to make final preparations when a powersat is to be completed. They will be transported and supplied by once-a-week flights of an Orbital Transfer Vehicle, a two-stage reusable rocket craft. This craft will burn hydrogen and oxygen brought from Earth.

For general day-to-day transport, there will be a network of enclosed buses running on tracks—a   **147**

A three-man work capsule uses manipulators to install a radiator panel. The capsule is at the end of a crane, which gives it a wide area of operation. Cutaway shows a construction worker guiding the manipulators while viewing the work through a bubble canopy. (Courtesy Boeing Aerospace Co.)

space rapid transit system, if you will, to carry people from the living centers to the far-flung reaches of the construction base. Each will carry two dozen commuters in shirtsleeve comfort. An airlock at the end of a cherry picker boom will allow the bus to pick up and deliver individuals to their enclosed control cabins on the construction equipment. Also, small free-flying spacecraft will carry inspectors or maintenance people to areas that otherwise would be inaccessible or hard to reach.

How long should people stay at their jobs? To establish the crew schedules, one can draw on expertise from Skylab, from nuclear submarines and undersea habitats, and from such activities as the construction of the Alaskan pipeline. On the Alaska pipeline, the most that people could stand to stay was eight weeks at a time. Similarly, the longest nuclear submarine missions are limited to seventy days; this is also the longest time that large crews have been confined in laboratory tests. However, medical data from the Skylab flights show that stays in space of up to one hundred days can be accepted. At Boeing studies of this problem suggest that the tours of duty should be ninety days. The work week would six days, with one day off per week (one wonders if schedules would be adjusted to make the day off fall on July 4 or Christmas). There would be two shifts, each working ten hours a day. The work day would be split into two four-hour sessions punctuated by half-hour breaks for lunch and dinner, followed by a two-hour evening session.

As on the Alaska pipeline, between their tours of duty the space workers will have time away from the project. They will be rotated back to Earth, their pockets bulging with their high wages, ready to enjoy ninety days off the job. Much of that time will be available for vacation, though some may get a few days of refresher training before returning to space. Yet there may be limits to how many tours of space duty a crew member can serve.

The crews will be exposed to radiation while in space. To diminish their exposure, most of the construction will take place at low orbital altitudes, where there is a good deal of protection from Earth's magnetic field. But there is danger from heavy atomic nuclei, such as iron, which stream in as high-energy cosmic rays. They can destroy cells, acting like hot needles as they pass through the body. Their effect on the brain is particularly damaging, since nerve tissue does not grow back when damaged. It has been calculated that on a three-year trip to Mars, some 10 percent of the brain's cells would be seriously damaged or destroyed.

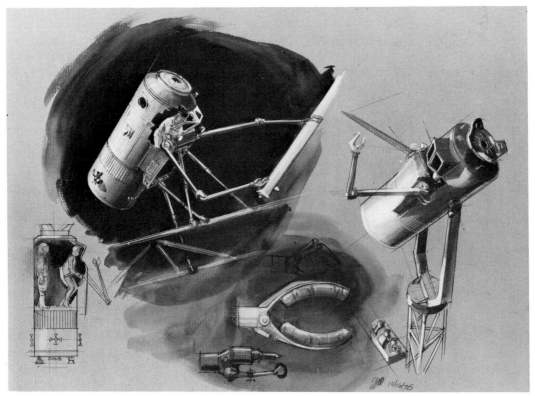

Tools for space construction. Upper left, *armlike manipulators of a one-man work capsule transport a beam. Other manipulators stablilize the capsule.* Lower left, *the worker in his capsule. Behind him is the pressure suit he uses when exposed to space.* Right, *the capsule mounted on a positioning crane.* Lower center, *tools used at the ends of manipulator arms.* (*Courtesy Boeing Aerospace Co.*)

Completed powersats are assembled in geosynchronous orbit. The construction base builds eight large solar arrays, two of which have transmitting antennas. These fly to geosynch on their own power using ion thrusters and are joined one by one in final assembly. (Courtesy Boeing Aerospace Co.)

There will be solar storm shelters in the living areas to provide radiation protection during the worst cosmic-ray outbursts from the Sun, but still the space workers will accumulate exposure to the point where they will have to be grounded. What may ground them at an even earlier date could be the effects of living in weightlessness. In zero-g, bones tend to lose calcium and to grow weaker. On the basis of results from the longest Skylab flight, eighty-four days, it appears that in one year of weightlessness a person would lose 25 percent of his bone calcium.

The body tends to replace its lost calcium when back in normal gravity, and indeed the Skylab evidence suggests that workers would replace their lost calcium during their ninety-day times on Earth between tours of duty. However, studies of middle-aged men undergoing prolonged bedrest have found that after six weeks in bed there is a tendency for the calcium replacement to go only part way and then level off. So it appears that long-term bedrest, and possibly space flight, causes permanent and cumulative damage to bones. There are drugs which promote the growth of bone tissue, but whether they will help remains to be seen.

In the whaling towns and seaports of the last century, it was not uncommon to find men who were prematurely aged, bent, gnarled, and dreadfully misshapen. They were victims of the bends: deep-sea divers who had come up too fast from under water, thus releasing bubbles of nitrogen from their blood and producing horrible agony. No one wants space workers to end up in similar straits, suffering from the peculiar diseases of their long-term flights. But if spacemen are to avoid such fates, it may become necessary to limit them to at most a year or so of total space flight in the course of their working careers. Should this happen, the resulting frequent turnover of people would drive up costs. There will be an obvious conflict between the two imperatives: "Protect the health of the people" and "get as much work done per worker as possible."

So there will be two trends that will shape developments once the power satellite program is well under way. The first will be to reduce further the number of people needed in space to build each powersat by bringing in more automation, better teleoperators, and robots. The rise of intelligent computers will greatly hasten this development, and may in fact be brought about by the demands of powersat construction. The second trend will be to provide longer staytimes and more allowable space exposure for those who indeed must live in space. This trend will lead to incorporating artificial gravity and better radiation shielding in work and living areas. There may also be attempts to grow food in space and to produce there the needed oxygen and water by closing the cycles of the environmental systems. With this, the habitations will advance in the direction of becoming true space colonies.

For all that robots and automation will grow in importance, an increasing demand for satellite solar power could well lead to a steady growth in the number of people needed in space, and hence to the sophistication of their living areas. It will be immaterial whether they think of themselves as astronauts, as orbiting work crews, or as space colonists. Specialists all, they will share the deep satisfactions of difficult work well done, of challenges faced and mastered. Whether their space careers be long or short, they all will feel the pride of achievement, of having lived through experiences shared by few. Their actual working careers will not be romantic, science-fiction idylls or space odysseys any more than the experience of building the Alaska pipeline was an exercise in communing with the northern wilderness. But they will share the camaraderie and fellowship of bold ventures, as in the days of Shakespeare's King Harry:

> God's will! I pray thee, wish not one man more;
> The fewer men, the greater share of honor.
> This day is called the feast of Crispian.
> He that shall live this day, and see old age,
> Will yearly on the vigil feast his neighbors.
> Then he'll remember with advantages
> What feats he did that day; then shall our names,
> Familiar in his mouth as household words,
> Be in their flowing cups freshly remembered.
> We few, we happy few, we band of brothers.
> And gentlemen in England now a-bed
> Shall think themselves accursed they were not here,
> And hold their manhood cheap whilst any speaks
> That stood with us upon Saint Crispin's Day.

> [*King Henry V*]

CHAPTER **8**

Paths of Commerce

For power satellites to be built, for a major reach into space to go forward, there are three things that have to happen first.

The project must, of course, be technically feasible, well-understood, with no insurmountable difficulties. It must be advantageous from the standpoint of economics, offering benefits at an acceptable cost and being competitive with other ways of getting energy. But there is a third requirement, and advocates of new space projects have often made the mistake of ignoring it or failing to understand it. This is the requirement of politics.

To many space advocates, politics is rather a black art, difficult to understand. For nearly two decades, space advocates have lived in the shadow of a single day, a unique and bold political act. The day was May 25, 1961, and the man was John F. Kennedy. It was on that day that he went before the Congress with a statement: "I believe the nation should commit itself to achieving the goal, before this decade is out, of landing a man on the Moon and returning him safely to Earth."

Everything that has happened since, all the dreams and hopes and aspirations, has flowed from that day. But Kennedy was no idle dreamer; he was a solid realist, and his decision was not made in a vacuum. It addressed the felt national needs of the day. It did this so strongly that not till the late 1960s, not till a time very different from the one Kennedy knew, did the decision arouse great controversy. It was not the magic of Kennedy and the charisma of his Camelot that built and sustained the Apollo program. It is worth remembering that while he was alive, Kennedy was unable to get Congress to pass such reforms as Medicare or his civil-rights bill. Apollo went forward because it had the support of the country. Not just of the space enthusiasts, nor the science fiction fans, but the nation.

As the 1980 presidential elections begin to approach, the country is once again on the verge of yet another political change. Like so much else in our contemporary culture, the movement started in California. At every statewide election it is the practice there to submit to the voters a set of numbered propositions, or proposed new laws. In the June 1978 primary election, Proposition 13 called for massive cuts in the property tax, as well as reforms in the way new taxes could be levied. It passed with 65 percent of the vote. Overnight, pundits everywhere were pointing to this as the initial act in a taxpayers' revolt, a movement toward cuts in government budgets and significant changes in patterns of taxing and spending.

Should this trend go forward strongly, the next few years will not be easy for those who seek new federal initiatives in space. The emphasis will be on budget cuts, on eliminating waste, on recognizing the limits of government action. But so far from destroying hope for the power satellite, the spirit of Proposition 13 may actually pave the way for its rise.

For when this movement has gained its successes and written its insights into law, among its most significant contributions may be a clear understanding of the things governments can and cannot do well. Among the latter would be military interventions, massive attempts at social change, and control of the economy. Among the former are traditional functions like defense and the courts and schools, collecting taxes and distributing payment checks—and, on occasion, mobilizing the nation's energies for major technical efforts.

The Panama Canal, the Manhattan project, the Apollo program—in ten years these may loom larger in the nation's memory than they do today. They may be seen as rare occasions when the federal government displayed qualities all too uncommon, yet all too valuable: competence, skill, daring vision, sound management, the ability to move forward rapidly and to focus in on national goals. At the same time, it will be obvious that there are problems of energy and of our world standing that the nation can and should address. Our position in the world will be seen to rest on the new ideas and new things we can provide. In a world in thrall to the energy of Arabian oil, there will be many who will look to a solution in the energy of America's creative minds.

Today in Washington the power satellite is unattractive to many officials because it could only be built as a massive federal program. Ten years hence, this defect may be a virtue: It may be seen as one of the few major federal initiatives that could address significant national needs while going forward on time and within budget. The strict fiscal discipline of past NASA and aerospace budgets then will tell; it will be a solid guarantee of performance. It will contrast with the financing of social programs, whose budgets by law have been declared ''uncontrollable.''

In the official Washington of 1990, with its trillion-dollar budget, the power satellite may be seen as a way to demonstrate the competence of government in meeting our energy needs. It may be hailed as a way of bypassing the difficulties of commercial fusion and the environmental problems of coal and nuclear plants. Its proponents will use Kennedy's phrase, that it will ''get this country moving again,'' while bringing forth the new generation of technical talent on which will rest our economic strength. Others will say that it would fire the country's imagination, lead us to look with hope to the new century rather than waste our strength in fruitless wranglings over the controversies of past years. Overseas, the export market may be justification enough, as the U.S. sells the works of its genius to balance its purchases of oil and raw materials. A world grown accustomed to U.S. communications satellites, computers, and aircraft may not long balk at getting power as well as data from American satellites. Nor may there be serious doubt as to the project's feasibility. With solid support from the technical community, with a public well familiar with the work of the space shuttle, with the continuing successes of a vigorous Soviet space program as inspiration, the power satellite may be accepted as naturally as was the interstate highway project in the 1950s. The ultimate presidential announcement may evoke a nationwide sigh of relief: "At last someone in Washington knows what he's doing."

Should there be such a national commitment, we will not want to piddle around half-heartedly with a handful of space shuttles. We will go forward to build the largest and best rockets we can conceive, just as with the Saturn in 1961. We will not be timid, nor will we shrink from the boldest of plans, but will seek the best means of addressing the question: How are the powersats to be built?

Even in an era of cheap space transport, it will be no mean feat to lift to orbit the megatons of propellant, materials, and supplies needed for a really serious powersat program. There is no doubt that at least the first few powersats will be built from Earth-launched materials. But once the powersats are actually operating, once the hopes of their advocates have been proven, there will be ample opportunity to seek lower costs and greater flexibility.

For all the awesome immensity of the Earth-launched powersat, it will appear in time to be quite conservative. There is a simple and straightforward line from Apollo and Skylab, through the shuttle and its early use in space construction, continuing on to more advanced shuttle-era projects built with the aid of a construction station, and on to the very large construction base to be used for production of full-size powersats. Even the long leap from shuttle-built communications platforms to true powersats introduces few truly new or untried procedures. It is more a matter of scale and organization, of operating many beam-builders and other equipment in a common program instead of merely a few.

Since an evolutionary approach can bring us so far, how much further can it take us? It is still too soon to answer this for certain. However, several years of study and analysis suggest that the ultimate answer will be "very far indeed" and that developments will proceed very much as has been foreseen by advocates of space colonization. These developments will arise out of the desire to lower the costs of powersats by building them from lunar resources.

This suggestion, that powersat costs might thus be reduced, originated with NASA official Jesco von Puttkamer in 1974 and gained the strong advocacy of Princeton's Gerard O'Neill. It was followed up by the economist Mark Hopkins at the 1975 Summer Study on space colonization and by Gerald Driggers at the 1976 Summer Study. But none of their work was truly convincing because their two powersat concepts, Earth-launched and lunar-derived, were designed and their costs estimated using entirely different ground rules. What was needed was a study that would look at the idea of a lunar-derived powersat with the same approaches and rules used for an Earth-launched design.

Precisely such a study got under way in April 1978 under the direction of Edward H. Bock at General Dynamics in San Diego. In contrast to the earlier Summer Studies of ten- and six- weeks' duration, this effort ran for ten months. It featured a seasoned team of professional engineers and consultants and had its overall management with the same NASA people who had managed Boeing's work on the Earth-launched powersat. These investigators thus were able to probe far more deeply than had the earlier studies.

Their first task was to design a powersat that could be built from lunar resources. This was a straightforward challenge; they took the Boeing design and broke it down to see what materials were to be used, then looked to see what lunar materials could be substituted. Their conclusion was that 91 percent of the mass of a powersat could come from the Moon. The remainder, some ten thousand tons per powersat, would consist of small quantities of silver, tungsten, and mercury, along with electronic components and other complex devices. Interestingly, in 1976 Gerald Driggers had also proposed that 91 percent of a powersat's materials could come from the Moon.

Next, it was necessary to understand how the materials were to be transported and where the processing plants to convert lunar soil into useful materials would be located. For comparison, "Concept A" was the standard Boeing scenario for Earth launch. "Concept B" was right out of the 1975 and '76 Summer Studies. It called for a small group of moon-miners to operate a mass-driver, flinging bags of soil into space, where a mass-catcher would await their arrival. From there, the mass would be shipped to a processing plant in high Earth orbit and chemically processed deep in space. The products of this processing plant then would serve in construction of powersats, which then would be transferred down to their permanent location in geosynchronous orbit.

The particular high orbit chosen for the manufacturing was the so-called "2:1 resonant orbit." There the facility would go around Earth twice while the Moon was going round once; hence, "2:1." In this orbit it would swing from some 100,000 to 200,000 miles from Earth in the course of its two-week period.

"Concept C" put the manufacturing operations on the Moon. Lunar soil would be processed there, with only finished industrial products launched to space. Transport would be by rocket. The

In contrast to Boeing, Rockwell's powersat proposal calls for most of the construction work to take place in geosynchronous orbit. (Courtesy Rockwell International Corp.)

rockets would burn hydrogen and oxygen: the oxygen obtained as a product of the lunar processing, the hydrogen brought from Earth. These rockets would not tarry at any intermediate orbit, but would proceed directly from the Moon to a construction base at geosynch. There they would serve the same functions as cargo rockets from Earth, delivering the needed supplies.

"Concept D" was quite similar except that the fuel for the rockets would be powdered aluminum rather than hydrogen brought from Earth. Powdered aluminum had the advantage of being available from lunar soil.

Any one of these approaches would crisscross space with paths of commerce, turning the unexplored regions of pre-Apollo days into well-traveled domains of human activity. Concept A, of course, would have only two regions for this activity: low Earth orbit with its construction base and geosynch for final powersat assembly and maintenance. The two lunar-rocket options, Concepts C and D, would add two more regions: the lunar surface and lunar orbit. The latter would be the site for a refueling depot for rockets passing to and from the Moon.

Concept B would be the most ambitious; it would have six locations for activity. There would be the four noted, plus the 2:1 resonant orbit for the manufacturing center, and the L_2 libration point for the mass-catcher. At this last location, a spacecraft can stay on station with little cost in propellant, since the centrifugal force of its orbital motion balances the combined gravitational pulls of Earth and Moon. L_2 is located forty thousand miles above the lunar far side.

How many people would live at these various locations, and how long would be their tours of duty? The four main locations are low Earth orbit (LEO), geosynchronous orbit (GEO), the Moon, and the 2:1 resonant orbit (2:1). Then, according to the General Dynamics study, there would be the following:

PEOPLE NEEDED IN SPACE FOR POWERSAT CONSTRUCTION

	Concept A (Earth-launched)	Concept B (Mass-driver)	Concepts C & D (Lunar rockets)
People at LEO × tours per year	480 × 4	60 × 4	60 × 4
People at GEO × tours per year	60 × 4	60 × 6	1,200 × 6
People on Moon × tours per year	—	60 × 2	400 × 2
People at 2:1 × tours per year	—	1,400 × 2	—
Total people per year	2,160	3,520	8,240

Many of these jobs are in the area of maintaining the complex automated facilities. The basic processing and production equipment were assumed to be heavily automated. In addition, industrial robots were included for materials handling, machine feeding, and machine unloading. A total of 1,651 robots was estimated for these routine production tasks, or 3.8 robots for each human operator.

These crews would build one powersat per year, each one producing ten million kilowatts for each of thirty years. This production rate may be lower than what might actually happen, but it gives an indication of what might be expected. Concept B calls for the fewer people because the lunar mass processing and powersat construction take place at the same location, and because the crew quarters can be heavily shielded against radiation, thus prolonging the staytimes. The shielding would be of lunar soil, which is much easier to transport to the 2:1 orbit than to geosynch. Since Concept B is straight out of the world of space colonization, this comparison scores a point in its favor.

An important item for comparison is the weight of cargo to be lifted from Earth. The "startup cargo'' establishes the needed construction facilities and industrial plants, as well as the crew quarters. The "steady-state cargo," so many thousands of tons per year, is needed for the on-going production. The General Dynamics study gave the comparison:

CARGO TRANSPORT (THOUSANDS OF TONS)

	Concept A	Concept B	Concept C	Concept D
Startup, 1,000's of tons	25.8	128.0	184.2	260.1
Steady state, per year	147.7	13.6	23.7	15.2
Total after thirty years	4,457	535	895	715

So there is another point on the side of the approach most favored by would-be colonizers of space.

Next comes the question of overall project costs. These costs would appear in several ways. There would be costs for research and development followed by program startup. In this phase there would be development and construction of the cargo rockets, as well as of the lunar or Earth-orbital industrial facilities. Then there would be production costs as the powersats appeared one by one. In addition, there would be costs for operating and maintaining the powersats, as well as the Earthside rectennas. The General Dynamics people gave cost estimates:

COST FOR THIRTY POWERSATS (BILLIONS OF 1977 DOLLARS)

	Concept A	Concept B	Concept C	Concept D
Research and startup	70.586	121.756	135.476	145.760
Production (one per year)	656.476	280.415	320.250	296.755
Operations and maintenance	186.651	186.651	186.651	186.651
Total for thirty years	913.713	588.822	642.377	629.166

The operations costs were taken to be the same no matter how a powersat came to be. This comparison scores a most telling point for Concept B.

However, the bottom line in this table shows costs the size of the national debt. To be sure, economists know only too well that such costs are indeed like the nation's electricity bill for thirty years. The question is, at what cost must electricity from powersats be sold in order to cover expenses, if the costs are to be computed in a way a utility company might find reasonable? Note the comparison:*

COST OF POWERSAT ELECTRICITY

	Concept A	Concept B	Concept C	Concept D
Electricity cost, 1977 cents per kilowatt-hour	5.74	3.40	3.78	3.69

These figures compare with costs of 3.5 to 4.3 cents per kilowatt-hour for future nuclear or coal-fired power plants, as estimated in 1978 by executives of Commonwealth Edison, Chicago's major utility company. According to *Science*, Edison Electric Institute reported 1977 figures for energy to be 1.5

*Startup and production costs are assumed to be divided evenly over the thirty powersats built one each year; each powersat is assumed to operate 95 percent of the time; and in each year of its operation, a powersat must generate revenues equal to 18 percent of its share of startup and production costs ("capital charge factor"). This last assumption follows standard practice in the utility industry. Operations and maintenance costs are figured on a pay-as-you-go basis. This method of cost estimating represents the views of the author and should not be regarded as reflecting the views of NASA or of its contractors.

Major elements of Rockwell's powersat would be extruded from a jig. (Courtesy Rockwell International Corp.)

cents per kilowatt-hour for nuclear energy, 1.8 cents for coal, and 3.7 cents for oil. In 1978 the figures were 1.5 cents for nuclear, 2 cents for coal, and 4 cents for oil. So the mass-driver approach scores one more point, the most significant of all, and wins the ball game.

Of course this is not the end of the matter, for ball games are played every day and few teams can win consistently. Similarly, one study does not a conclusion make, and many more studies will be in order before we will really be sure we understand the best way to build powersats. For instance, it is possible that this conclusion was influenced by internal NASA politics and the fact that the study was managed by the Johnson Space Center and not the Marshall Space Center.

NASA-Johnson and NASA-Marshall are two of the largest space centers, and their people have done yeoman work in the Apollo, Skylab, and Shuttle projects. In recent years their senior managements have increasingly felt that their centers' futures would depend on getting a large share of the action in any power satellite effort. Thus, NASA management has not given responsibility for powersat studies to any one center, but has allowed each center to pursue its particular design approach. So it was, in 1977, that Johnson contracted with Boeing and Marshall with Rockwell International Corporation to carry out major studies.

The Boeing approach, discussed in Chapter 7, called for the powersat to be built in the shape of a single flat slab with transmitting antennas at each end. Power would be generated by silicon solar cells. The principal construction operations would be in low Earth orbit, where the construction base would build each powersat in eight sections resembling the leaves of a dining-room table. Each section (two of them would carry antennas) then would be fitted with ion-electric rocket engines and fly under its own power to geosynch. The ion engines would use electricity to eject atoms of argon at very high speeds, some 225,000 feet per second, to produce thrust.

Activities at geosynch would be strictly limited. Because each powersat section can produce much more power than it needs for the electric rockets, many of its solar arrays would be rolled up like window shades. The few crew members at geosynch would unfurl the arrays, causing the powersat sections to spread sail like a clipper ship. As each section arrived, at forty-day intervals, it would be joined to the others. A completed powersat would be activated by a ground station.

In contrast, Rockwell's design called for the use of huge mirror reflectors that concentrate sunlight on the solar cells. The reflectors would not be of silicon, but of a different material, gallium arsenide. The most unique feature of Rockwell's approach was that the main construction would take place in geosynch. The largest work crews would live there, and the powersat would be built as a single unit rather than piecemeal.

Both NASA-Johnson and NASA-Marshall agreed on the need for a transfer rocket, a Personnel Orbit Transfer Vehicle, to take people and their supplies to and from geosynch. It would be powered by hydrogen and oxygen brought from Earth. To get cargo from the Earth to low orbit, both centers agreed on the need for the huge cargo rockets, though they differed as to the best designs. For

Personnel Orbit Transfer Vehicle (POTV) would be a two-stage craft, both stages to be fully reusable. It would transport people and priority cargo from low orbit to geosynch. (Courtesy Rockwell International Corp.)

Cargo Orbit Transfer Vehicle (COTV) would resemble a huge box kite. Fitted with solar arrays and ion thrusters, it would leisurely transport cargoes of 10,000 tons or more from low orbit to geosynch. (Courtesy Rockwell International Corp.)

CHEMICAL PROCESS PLANT CONCEPT (APPROXIMATELY TO SCALE)

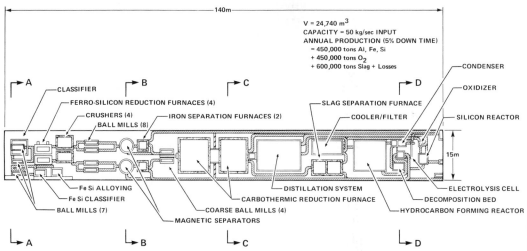

Development of a lunar resources processing plant would begin in low Earth orbit. Here a test plant prepares to prove out the techniques for obtaining aluminum, glass, oxygen, silicon, and iron from lunar soil. Its crew live in the crossed barbell-shaped structures, which rotate for artificial gravity. (Courtesy Rockwell International Corp.)

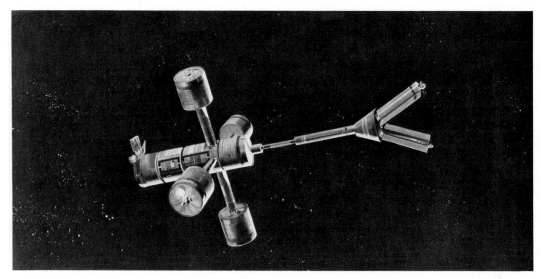

Detail of the chemical processing plant. It would be built up alongside the orbital construction base. Then base and plant together would use the COTV for transfer to the high orbit at which lunar resources are most accessible. (Courtesy Ames Research Center)

transporting cargo from low orbit to geosynch, the Marshall designers proposed an immense solar-powered Cargo Orbit Transfer Vehicle. Looking like a huge box kite, it would be fitted with solar panels and ion-electric engines to shuttle slowly and majestically between the two orbits. The need for several hundred workers in geosynch, rather than just a few dozen, turned out not to be a major problem; it would not significantly increase costs.

Clearly, no serious choice could be made between these approaches, or one of the NASA centers and its contractor would be left out. There was an attempt to come up with a powersat that would combine the best features of both. It ended up by assigning responsibility for major items of effort between Johnson and Marshall, in the fashion of eenie-meenie-miney-mo, or of ''one for you, one for me.'' The resulting hybrid was called the ''camel,'' a camel being defined as a horse designed by a committee. Few people within NASA took it seriously, and the likelihood was that Johnson and Marshall would pursue their approaches more or less independently.

If the powersat construction were to take place in geosynch, then it could be easier to bring in lunar materials by means of rockets from the Moon. There would be the need for an elaborate lunar base capable of producing rocket fuel and of turning lunar soil into industrial goods. But in the transition from Earth resources to lunar, life at the geosynch construction base would go on as before. The main difference would be that the cargo rockets would arrive from the Moon, not the Earth. This indeed is the scenario which Ed Bock would have studied if his funding had come from Marshall and not Johnson. It could have meant a decided fillip to Concept C or D.

However, this proposal really is not at all likely. At eighty-five hundred tons, the Boeing construction base would be scarcely heavier than a section of the powersats it would build, so that it could easily be moved to higher orbit. By spreading solar arrays like sails, and by installing ion drive, this base would be only slightly more difficult to fly to the 2:1 orbit than to geosynch. It could then **161**

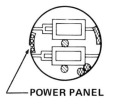

POWER PANEL

A-A

B-B

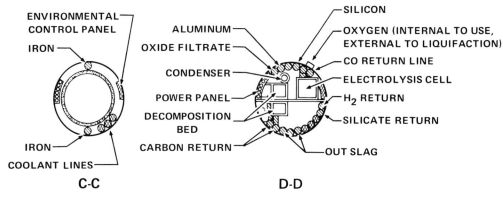

ENVIRONMENTAL CONTROL PANEL

IRON

IRON

COOLANT LINES

C-C

ALUMINUM

OXIDE FILTRATE

CONDENSER

POWER PANEL

DECOMPOSITION BED

CARBON RETURN

SILICON

OXYGEN (INTERNAL TO USE, EXTERNAL TO LIQUIFACTION)

CO RETURN LINE

ELECTROLYSIS CELL

H_2 RETURN

SILICATE RETURN

OUT SLAG

D-D

Cross sections of the chemical plant. (Courtesy Ames Research Center)

PROCESSING OF LUNAR SOIL

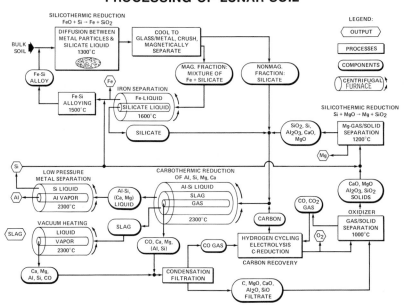

Schematic drawing of the chemical processes proposed for use. (Courtesy Ames Research Center)

serve as the construction site for powersats, as in Concept B. Even more important, while still in low orbit it could be fitted out with its materials-processing equipment. It will be quite important to assure the proper functioning of this chemical processing plant in space, and everyone concerned will breathe easier if help is close at hand when things go wrong. And low Earth orbit is a lot closer than the Moon.

So this would continue the cautious, conservative, step-by-step approach. Bit by bit, the low-orbit construction base would receive its processing plant equipment, and with the necessary tests and checkouts, its reliable operation would be assured. It would not process lunar soil at first, but would be tested using lunar-like sand brought from Earth. One wonders what some senators will say when they learn that a huge cargo rocket has lifted off from Cape Canaveral carrying a five-hundred-ton cargo of crushed gravel.

Several hundred people will work at the processing plant, which will be highly automated. As far as possible, the processing will involve continuous flow, and the plant will be set up to run like a modern oil refinery or chemical company. The comparison is instructive. In December 1978 some sixty thousand oil-refinery workers walked off the job to strike for higher pay. But the refineries were so highly automated that they easily kept running under the control of the strikers' supervisors. In a long strike refinery equipment would have suffered for want of maintenance. But the strikers, finding their absence made little immediate difference, soon went back to work.

Once there is a fully functioning manufacturing and construction base in the 2:1 resonant orbit, it will be a latter-day version of Detroit's River Rouge plant, which takes in raw iron ore and turns out complete autos, all at one location. The space center will receive unprocessed lunar soil at one end and turn out a finished powersat at the other, ready to fly on ion engines down to geosynch.

A powersat takes shape in high Earth orbit, less than 100,000 miles from the Moon. Left, *people live in shielded modules that resemble a pinwheel and rotate for gravity.* Center, *a docking station for transport rockets.* Foreground, *a small free-flying capsule with a worker inside. Solar panels are being installed at bottom and right. (Courtesy General Dynamics)*

The step-by-step completion of this center will finish what by far will prove to be the most difficult job. The remaining problems will mostly be solved on the Moon or near it. By comparison with the powersat project, and with the development of the manufacturing center, these efforts will seem minor. They will have much more in common with some of the more ambitious lunar-exploration programs proposed for the post-Apollo era of the 1970s. These efforts will provide for the actual shipping of lunar resources from the Moon to the 2:1 center; and in the most difficult parts of the problem there are solutions that are not only feasible but downright elegant.

The first of these problems involves the sensitivities at launch. The mass-driver acts like a cannon or catapult, and cannon often miss their targets. The payloads to be launched will be mere bags of soil, flying without guidance or course correction once away from the Moon. The target they must hit is the mass-catcher, forty thousand miles away near the L_2 point, yet only three hundred feet wide. This is such a difficult problem in aiming that if the payloads are launched with a velocity error of only a few

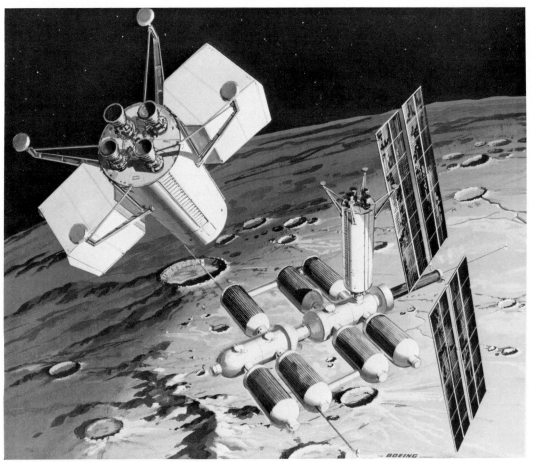

Cargo handling in lunar orbit. A POTV from Earth, docked to the orbiting lunar station, has transferred its cargo to the lunar landing vehicle in the foreground, which is departing for the Moon. (Courtesy Johnson Space Center)

Cargo rocket lands on the lunar surface. (Courtesy Johnson Space Center)

microns per second (the speed of a swimming bacillus), they would be expected to miss. Few marksmen have ever faced such a problem.

The problem was studied during a 1976 summer study on space colonization of which I was a part and which was headed by Brian O'Leary, a former astronaut. The solution used my discovery of a new effect that amounted to a kind of focusing. Payload flight paths making use of this effect came to be known as "achromatic."* Using achromatic trajectories, there could be a launch velocity error of as much as ten centimeters per second, which would be easy to achieve. The payload would have a miss distance at L_2 of only twenty meters. Over the next year and a half, while I held a research fellowship in Germany, I went on to study these achromatic trajectories in detail. It turned out that this effect led to a complete understanding of the problem of lunar mass transport. In particular, the mass-catcher could always maneuver so as to be reachable via achromatic trajectories.

The mass-catcher represented the second major problem. The earliest suggestion for its operation dates to a NASA-Marshall report of January 1975, "Space Colonization by the Year 2000: An Assessment." That report pointed out that even if materials were launched from the Moon by mass-driver, it would be no mean feat to catch them:

---

*In a camera, achromatic lenses bring light of different wavelengths to a common focus. Similarly, achromatic trajectories bring payloads launched with different velocities to a common target.

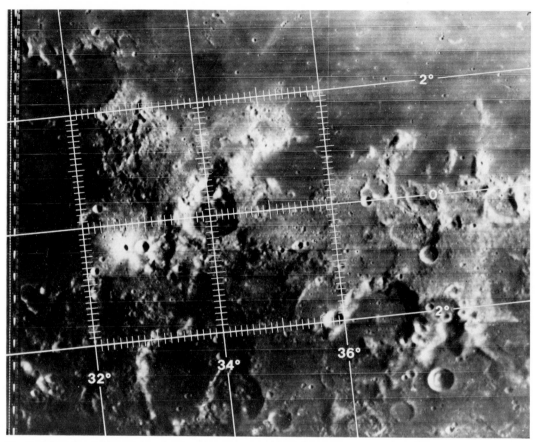

The lunar base is to be located near the crater Censorinus (bright spot at lower left). Two very good locations are around 2° north by 33.3° east, and 0.7° north by 34.8° east. (Courtesy Ames Research Center)

There is nothing simple about catching [the stream of lunar rock]! . . . As ridiculous as it sounds, the best mechanism that could be devised in this analysis is a large funnel with reaction engines (possibly electric ion engines supplied with solar power). And even then the aim from the lunar surface must be extraordinary.

Without being aware of the Marshall work, at the 1975 Summer Study on space colonization I also suggested that there should be a mass-catcher and recommended it be in the shape of a large cone-shaped bag. But what was not understood was how payloads would actually be captured or caught when they hit the bag. They might just bounce out and be lost.

Help came from an unexpected quarter. In Tucson, Arizona, astronomer William Hartmann was studying what on the surface would look like a totally different problem: the origin of the planets. For years, planetary scientists had agreed that the planets must have been built up from many tiny rocky bodies, or planetesimals, which came together and stuck. But they could not understand how such bodies could have stuck together when colliding. It was much easier to believe they would have rebounded from each other like billiard balls.

In his laboratory Hartmann found that this assumption was not so. Such colliding bodies would tend to chip and fragment, forming quantities of dust or sand, which would cling to their surfaces. The layer of dust, called a regolith, would act as a shock absorber. It would absorb the impact of another body, and the feeble gravity of small planetesimals then could hold them together. Hartmann found that even thin layers of regolith would be quite adequate as shock absorbers.

It was easy to see how the same thing could happen in a mass-catcher. Bags of lunar soil would split open when they hit the catcher, and their contents would form a new regolith. When other bags hit, their impacts would be absorbed. Then a modest rotation of the bag would provide artificial gravity to hold everything in place.

From there it would be easy. When the catcher became full, it would accelerate forward while slowing down the spin of the bag, till its contents collected in a compact mass. The caught mass could then be packaged by the simple technique of pulling drawstrings to enclose it in a flexible container, as if it were so much dust collected in the bag of a vacuum cleaner. The catcher then would back away, releasing this large cargo into free flight. It would loop round the Moon, then escape into a high orbit of Earth.

Detail of the lunar base. Top left, *a mine, where loose soil is scooped up like sand at a beach.* Top center, *the mass-driver: an electromagnetic catapult that launches bags of soil to be caught by a mass-catcher. Base structures are covered with soil for protection against radiation and temperature extremes. (Courtesy Johnson Space Center)*

Payloads launched from the mass-driver pass through a downrange fine-guidance station, which corrects their aim and ensures they will hit the catcher. (Courtesy Ames Research Center)

The cargo then would go to the space manufacturing center in 2:1 resonant orbit. While I was consulting for Brian O'Leary in 1976, I also was wondering about this problem and found that the 2:1 orbit was the stable orbit most convenient to reach in this way. Of course, the cargo would not go directly to the center, but would merely pass in its vicinity. So a rocket-powered craft known as the terminal tug would be based at the center to go out and retrieve the large bag of lunar soil.

The bags could well run to a hundred thousand tons—the capacity of a seagoing ore carrier. It seems absurd to speak of handling such loads with mere rockets, but it's not. The cargoes will pass so slowly by the center that retrieving them will call for propellant tanks of quite modest dimensions: twenty meters diameter for the hydrogen, fifteen meters for the oxygen. A single flight of one of the large cargo rockets will carry the needed hydrogen from Earth to orbit.

The mass-catcher will be a most curious type of hybrid since it will have two distinctly different propulsion systems. Its hydrogen-oxygen rockets will take it from the 2:1 center to L_2, as well as permit it to maneuver near the 2:1 center in order to retrieve the cargoes. These rockets will also serve for the "ullage" maneuver, which compacts caught mass in the catcher bag so that it can be packaged. For the continuous maneuvering near L_2 during catching operations, it will have a second system: an electric propulsion drive. The power plant will almost surely be nuclear, since it is a safe prediction

Launch of payloads from the Moon. Lunar base (with mass-driver) is at right. The payloads form a stream which arcs across the lunar surface and then rises beyond the horizon (left), *taking two days to reach the mass-catcher. (Courtesy Don Dixon)*

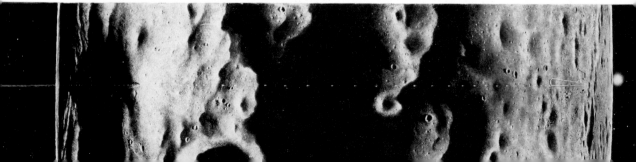

that some payloads will miss the catcher. These would rip solar panels to shreds, but a nuclear reactor can be protected from damage.

With this, the roster of spacecraft and systems is complete. It is a rather impressive list:

Cargo rockets, to carry some five hundred tons of payload from Earth to low orbit.

An advanced version of the space shuttle to carry seventy-five people to and from low Earth orbit.

Personnel Orbit Transfer Vehicles to carry people as well as small priority cargoes. These provide connecting service between low Earth orbit, geosynch, the 2:1 orbit, and a low orbit of the Moon.

Cargo Orbit Transfer Vehicles, solar-powered and with ion drive, to carry large cargoes from low Earth orbit to the 2:1. These may also tow complete powersats from the 2:1 to geosynch on the return leg.

Space propellant depots, in low Earth orbit, geosynch, low lunar orbit, and at the 2:1. They will be capable of liquefying oxygen and perhaps hydrogen, so that these propellants can be stored indefinitely.

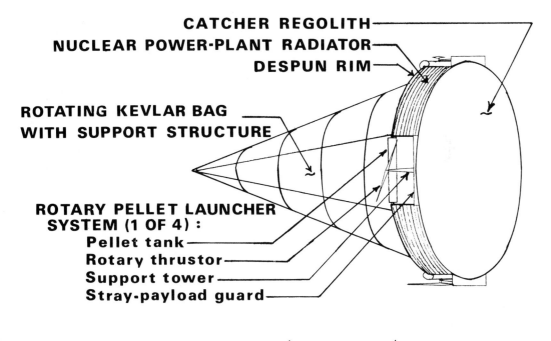

CATCHER REGOLITH
NUCLEAR POWER-PLANT RADIATOR
DESPUN RIM

ROTATING KEVLAR BAG
WITH SUPPORT STRUCTURE

ROTARY PELLET LAUNCHER
SYSTEM (1 OF 4):
Pellet tank
Rotary thrustor
Support tower
Stray-payload guard

|—————— 50 meters ——————|

The mass-catcher. Payloads fly in from the right and break up to form a layer of loose soil (a regolith), which aids in capture of subsequent payloads. Catcher is 40,000 miles behind the Moon, but payloads fly in with high accuracy due to their use of achromatic trajectories. (Drawing by the author)

A maintenance and repair center for the powersats in geosynch.

The space processing and manufacturing center at the 2:1 orbit.

Lunar ferry rockets, to shuttle between an orbiting lunar station and the site of the mass-driver.

The mass-driver and lunar base, with a mine for lunar soil.

The mass-catcher.

And, not to be overlooked, the powersats. Most of them will be in geosynch.

Most of these facilities will call for space crews. There will be groups of perhaps a few hundred people living in the powersat maintenance center in geosynch, shuttling to and from the immense craft that will be in their care. A few dozen will live at such lonely outposts as the Moon and the four refueling depots. The true centers of the project will be the locales having the largest numbers of people: geosynch and the 2:1. Studies at Boeing propose that the powersat maintenance center have

A space refueling depot. Propellants are stored in the numbered modules. At left a POTV tanks up for a flight to the Moon. Depot has facilities for maintaining propellants in a liquid state. Similar depots exist in geosynch, in the 2:1 resonant orbit of the main manufacturing center, and in lunar orbit. (Courtesy General Dynamics)

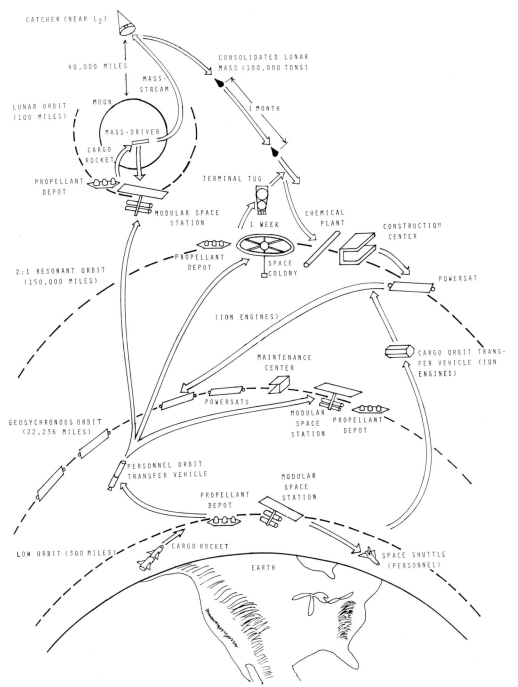

CATCHER (NEAR L₂)

CONSOLIDATED LUNAR
MASS (100,000 TONS)

40,000 MILES

MASS-
STREAM

1 MONTH

LUNAR ORBIT
(100 MILES)

MOON

MASS-DRIVER

CARGO
ROCKET

PROPELLANT
DEPOT

TERMINAL TUG

CHEMICAL
PLANT

CONSTRUCTION
CENTER

MODULAR SPACE
STATION

1 WEEK

POWERSAT

2:1 RESONANT ORBIT
(150,000 MILES)

PROPELLANT
DEPOT

SPACE
COLONY

(ION ENGINES)

CARGO ORBIT TRANS-
FER VEHICLE (ION
ENGINES)

MAINTENANCE
CENTER

POWERSATS

MODULAR
SPACE
STATION

PROPELLANT
DEPOT

GEOSYCHRONOUS ORBIT
(22,236 MILES)

PERSONNEL ORBIT
TRANSFER VEHICLE

MODULAR
SPACE
STATION

PROPELLANT
DEPOT

LOW ORBIT (300 MILES)

CARGO ROCKET

EARTH

SPACE SHUTTLE
(PERSONNEL)

Paths of commerce. Main centers of activity are in low Earth orbit, geosynch, the 2:1 resonant orbit, lunar orbit, the L₂ point (40,000 miles behind the Moon), and the lunar surface. (Drawing by the author)

Small modular space station of the type that will support the cargo-transfer and propellant depots. Similar modules will serve initially for housing the work force and will resemble those of Chapter 7. (Courtesy Rockwell International Corp.)

480 people on ninety-day tours of duty. Such a work force could perform needed services for forty powersats each year. At the 2:1 center would be the 1,400 people on six-month tours.

Wherever there would be need for space crews, there would be crew modules to live in. The project managers will like such modules, perhaps more than will their inhabitants, because they are convenient. They can be built, fitted out, carefully tested, and proved on Earth. Then it is a straightforward matter to lift them to space. Even large one-hundred-person units fifty-five feet in diameter will fit neatly into the cargo bay of a single cargo rocket. While such modules will have far from identical internal layouts and systems, there will be enough similarity to make them fairly easy to maintain. They will readily be grouped in clusters, and if more are needed it will be simple to bring them up.

In the small stations at low Earth orbit and low lunar orbit, as well as in geosynch, the modules will be of the simple type described in the last chapter. Quite likely they will lack both radiation shielding and artificial gravity, at least for a while. On the Moon the modules will have coverings of lunar soil for protection both against radiation and nightfall: Outside their snug security, temperatures will rise and fall through a range of four hundred degrees. The most elaborate modules will be at the 2:1 center. There it will be possible to surround them with thick coverings of lunar sand for radiation protection. Also, like weights at the end of a barbell, modules will be rotated in pairs for much-desired gravity.

For all that, life at the 2:1 center will still be a far cry from the golden luxuries predicted for the Stanford torus space colony. The brief tours of duty and frequent crew rotations, the dependence upon supply flights from Earth, the cramped submarine-like surroundings will all be much more like Skylab. At that point, space colony buffs may be forgiven if they feel their dreams are as far as ever from realization.

Yet it is always darkest before the dawn. It is precisely then that the influences that can usher in the first true space colony will begin to grow.

By providing shielding and artificial gravity, people will be able to serve many tours of space duty without harming their health. Yet few will want to. The recreation opportunities will be limited, the artificiality of the surroundings overwhelming, and the chance for normal family life virtually nil. Life in the modules, service in the powersat project will be very much an adventure for the young. There will be many who will seek to spend a few years of their careers this way, collecting their high salaries while serving a few tours of duty in space. Then, in the words of a Continental Army recruiting poster of 1776, "he may, if he pleafes return home to his friends, with his pockets FULL of money and his head COVERED with laurels."

The transient nature of space crews will create the need for a huge continuing investment in recruiting and training. As the project grows and expands in space, the Earthside training centers will

The first space colony may be built to provide a homelike atmosphere for the people working in space. The chance for normal family life in space would encourage many good people to stay with the powersat program. (Courtesy Ames Research Center)

Detail of the center of the Stanford torus colony. Spacecraft can dock at the central hub. (Courtesy Ames Research Center)

have to develop apace. There will be an enormous waste of talent and experience as people come and go without staying long. Many good people will leave the project just as they are at the point of their most significant contributions. Worse, a shortage of good people may actually delay needed expansion and growth.

It bids fair to imagine that after a few years the glamour and excitement of being a space worker will lose some of its attraction. Potential employees may look on space as merely one more of a variety of interesting and challenging ways to spend part of their careers. There may also be a problem if many space jobs call for skills and types of work not easily transferred back to Earth. The project veterans then could face the unpleasant choice of remaining with a space job that has become an inconvenience or risking unemployment by leaving. This possibility will become especially problematic as space jobs grow increasingly complex and specialized. There may be an echo of the early-1970s recession in the aerospace industry, when veteran employees often were told they were ''overqualified'' when they sought work outside that industry.

As any senior powersat manager will say, the most important resource will not be the metals of the Moon or even the robots and other complex equipment. The most important resource will be the project's people, and it will be quite important to keep the good ones with the project. As with any other effort calling for highly trained and skilled people, it will be very important to develop a solid core of career-minded individuals willing to live in space and to make it their home. This will bring other benefits as well. Crew members at the lunar base or at the geosynch maintenance center or refueling outposts will work better if they regard space and not Earth as their home. In the same way that many sailors regard the sea as home, no training, no mere experience of a few years can match the sense of ease and familiarity that comes from regarding space as a natural part of one's entire life.

There are two possible ways to solve these personnel problems, and probably both will be used. The first is to set up a space academy. It would offer full scholarships to selected space cadets in a

Interior of the Stanford torus. (Courtesy Ames Research Center)

Though somewhat crowded, the interior layout of a space colony could resemble that of some old European towns. Nordlingen, in Bavaria, could be thought of as an Earthside space colony; its shape, dimensions, and population are similar to those of the Stanford torus. People have lived for centuries behind the medieval wall. Yet Nordlingen has a large church, parks, streets, and mostly single-family houses. (Courtesy German Federal Railroads)

program leading to the degree of Bachelor of Science. In return, the students would agree to spend the first five years of their careers with the space program. This, of course, is nothing more than what we have today in the Coast Guard and merchant marine academies and in the three academies for the military services.

It is an instructive and mind-stretching exercise to imagine what the U. S. Space Academy would be like. How strong would be the emphasis on the liberal arts and on traditional subjects like astronautics, electronics, robotics? How will this school resemble or differ from the Air Force Academy or Annapolis? Will it field a football team and how good will it be?

The Space Academy would be very much on Earth, and doubtless many a congressional district will vie for its location. The same will not be true of the second approach to those personnel problems. This is the space colony.

The first space colony may have the shape and internal layout of the Stanford torus described in Chapter 3. Yet it will be but a step toward the goal of true space colonization. If the crew modules will be like quarters in a submarine or offshore oil rig, the first space colony will be like living aboard a supertanker. In today's European merchant marines it is quite common even for junior officers to have large and well-appointed cabins with carpeting and comfortable furnishings. Frequently they take their wives along, and an approximation to normal family life is by no means unknown. A few years ago, Globtik Tankers, Ltd. went so far as to announce plans to have a school aboard one of their tankers, so that families could stay there year-round. Nor do the womenfolk spend their days looking out the porthole while waiting for hubby to come home from the engine room. On the European ships they are treated as seagoing cadets and assigned a variety of useful tasks.

So it will be when the first space colony is opened for settlement. With the Space Spider and the other construction techniques available, building it will prove straightforward, perhaps even a trifle anticlimactic. At first the people will know only (only!) that they will have much more room, greater comfort; not mere cubicles, but true homes to call their own. But they will still be dependent at first on Earth for their food and other needs, just as before. Community life will grow slowly. The all-important systems for environmental control may simply be transplants from the old crew modules.

These families, these embryonic space communities will truly be homesteaders on the space frontier. Slowly, and with many a false start, they will develop the arts of space agriculture. They will close their environmental cycles, recycling what is needed, and in time their dependence on Earth will lessen. In contrast to the cramped crew modules, from the first they will break with a space tradition dating to the earliest Vostok and Mercury capsules. They will have room enough, and more than enough; for the Stanford torus will be too large for the initial small band of no more than a few thousand. In time their numbers will grow and they will fill to its limits; but by then the next colony will be ready.

By linking several Stanford torus-type colonies together it would become possible to develop the arts of space agriculture. (Courtesy Ames Research Center)

By the year 2050, give or take a few decades, all of this may come to pass. Even before the end of the present century it is quite reasonable to say that the key problem of space transportation will be solved or well on its way to solution. A century of effort in astronautics will at last culminate in standard, reliable rockets that can serve all needs. Thereafter, progress in rockets will continue, but only in the sense that there are advances in the design of automobiles and ships. The really revolutionary developments will lie in the past.

The same will be true of the techniques of space construction. Even today we can look ahead confidently to the automated equipment that will serve the building of powersats. Only slightly further ahead lie true robots and chemical plants to process material from the Moon. Continued progress in electronics and computers, and in the building of plants to extract metals from low-grade ores, will provide the solid underpinning here. As with jet aircraft, space construction will cease to be an experiment and will instead become an industry.

What may take longest to develop will be solutions to the problem of providing for thousands of people to live in space amid self-sufficiency. Space workers will be so productive, and the value of their work so great, that even with a fully developed powersat program there may never be need for more than ten thousand people to live in space. The importance of their contributions to Earth's economy will be enormous, quite sufficient to justify the cost of creature comforts made on Earth. Even so, there will be ample reason to seek to develop space agriculture and to close the loops in their life-support systems to permit recycling of materials.* A population of ten thousand can be supplied by rocket, but the effort would amount to virtually a continuous Berlin Airlift. That airlift succeeded in supplying Berlin under a Soviet blockade in 1948–49, but the effort proved quite taxing to our Air Force. Surely the presence of large space populations will stimulate much attention to promoting their independence from such large-scale resupply and to producing from space resources as much as possible of what people will need to live in comfort.

In the colony they will grow gardens and set up parks where they can pursue their friendships and loves. In a decade or so, Earth will be like Europe to many in the U.S.—a fascinating and richly rewarding place to visit, but not a place to live. Space will have lost its terrors, even its inconveniences; it will be comfortable and familiar. It will be the place where people have their jobs, their homes, their school, their lives. Space will no more be a place merely for the venturesome, for the explorers and pioneers. It will be home.

---

*Descriptions of space agriculture and closed-cycle space ecological systems, as well of the architecture and furnishing of homes in a space colony, are in my earlier book *Colonies in Space*.

CHAPTER 9

The Orbiting Bureaucracy

There is a popular game among the people who are familiar with space colonization. It is a game that is a lot of fun to play, for there are no rules. What's more, any number can participate. The game is to design a structure for a space colony society. The goal is to decide how the space colonists should live.

This game is actually a kind of intellectual Rorschach test. That is, the topic is one on which there is little available information or knowledge, yet one which tends to bring out a lot of intense emotional reaction. Often the reactions tell us very little about the subject of space colonies, but they tell us quite a lot about the people who make the social proposals. The range of possibilities that have been suggested is really quite vast.

One proposal often heard is that space communities should be under a quasi-military rule. This idea can be made superficially plausible by appeal to the presumed dangers lurking about, the delicate balances which allegedly must be maintained, and the supposedly ever-present threat from the deviant, the social nonconformist. What's more, there is the appeal of what purports to be a historical justification: Were not all the astronauts military men?

The answer is: No, they were not. True, many of them had military backgrounds, but none wore their uniforms into space. As astronauts, they were not under military discipline or subject to the Uniform Code of Military Justice. They were disciplined, true, but theirs was the self-discipline of highly motivated individuals. In any case, we cannot say that because the early space explorations relied upon close adherence to a code of regulations such regulations should be part and parcel of the human future in space. One may as well say because Captain James Cook enforced British naval discipline aboard the *Endeavour,* that therefore in the lands he discovered—Australia, New Zealand, Hawaii—only a military dictatorship will do.

179

On the leftward side of the political fence, some writers have let their imaginations run free in envisioning some form of Marxism or social utopianism. They have raised the argument: Space communities will offer opportunities for new social forms, so why not start right off with these forms?

Certainly it is true that in the long run space communities may offer important new concepts to the world at large—new political organizations, new social forms, new and more satisfying types of community organization. In just this fashion did America develop the concept of democracy as a new form of government in the world. But it will be quite important that such new forms arise out of the felt experiences and needs of the space-dwellers themselves. They should not be imported, for if new social forms have not taken root on Earth, why should we think they would do so in space?

Nevertheless, let us think about the opposite viewpoint. Suppose that through a good deal of intelligent foresight and planning, a new social form (nature specified) is introduced into a space colony and is accepted there. Would it then follow that humanity had made an advance, that mankind had by this experience learned something new and useful? It would not.

For one thing, the selection system would favor those people predisposed to the system. Opponents of the new social form would avoid emigrating or else would raise objections in their pre-selection interviews, so that they would tend not to be chosen. On the other hand, advocates of the new forms would be all the more attracted (and attractive) to the space communities. Therefore the problems of making the system work with a group that had some people opposed to it would be avoided.

In addition, there is the "Hawthorne effect." It is a social effect that involves groups of people who are made to feel they have been selected for special attention. It takes its name from an attempt to improve the output of workers in a telephone factory in Hawthorne, Illinois, beginning in 1927.

In that factory the management was concerned with arranging the working conditions so that it would be easier for the people there to turn out a lot of telephones. The managers hoped to find how to change the lighting, the arrangement of the work stations, even to a degree the nature of the employees' responsibilities. They expected that some of the changes would make things easier while some would make things harder for the people; but that's not what they found. Regardless of what they did, the workers' output went up.

It turned out that what mattered was not so much the levels of lighting or the other arrangements. What mattered was that the people felt they had been singled out by management to receive special attention. They were quite flattered, so they did better work. The stroking of their egos led to the increases in their output.

A space community would be a Hawthorne experiment par excellence. The space residents would from the start be well aware of their special roles. They would gladly put up with, even learn to enjoy, a wide range of situations that ordinarily they would avoid or come to dislike. So it is hard to see how a space community indeed would be a valid laboratory where new social forms could be tried.

There is one other type of social structure frequently proposed that actually takes in not only the social structure but also the overall organization of the project. Proposed is that the space community be international, that it be settled by a variety of people from different countries rather than being restricted to the U.S.

In thinking about this, it is important to understand the nature of the interest foreign nations would have in a space community and its products. There are several international treaties in force that would govern such a project as building power satellites. These treaties do not require that even the largest space projects be international. They simply require nations to refrain from laying claim to regions of space or of celestial bodies, after the fashion of the conquistadores. This restriction should not be a problem.

But if a space community is to build powersats for the world, then the world's nations will want to see an international authority exercising control over the powersats. It would prevent arbitrary changes in the rates or political decisions favoring one nation over another. Such arrangements will not be hard to set up, and in fact will likely follow existing practice in satellite communications and international aviation.

The Intelsat consortium operates the world's satellite communications system. It was first set up in 1964; by 1975 it had grown to eighty-nine member nations. Every thirty days it sends out its bills to each of these nations. It controls rates, standards of service, access to channels on the satellites. Significantly, the U.S. does not manage or control Intelsat, but furnishes technical advice under contract.

On the other hand, Intelsat does not build the communications satellites. Instead, it contracts with space companies like Hughes or Lockheed for them. The satellites themselves then are built by Americans, but are paid for by Intelsat. They are launched using NASA's rockets, but Intelsat then reimburses NASA for the launch costs.

In aviation it is quite similar. Outside the Soviet bloc, most commercial airliners are built by Boeing, Lockheed, or McDonnell Douglas, which are American firms. Yet nobody demands that the work forces there be international or that these companies be internationally owned. But international aviation, per se, is under the control of IATA, the International Air Transport Association. As the rate-making body, it determines the fares and the conditions for international airline service, and this body *is* international.

So it is hardly necessary that space communities be under United Nations control, or that their populations be drawn from many different countries. Nor is it necessary that the crews of the lunar base resemble those of a Liberian-registry freighter. It is quite reasonable to think of a power satellite project as a purely U.S. effort, with some sort of international jurisdiction over the export market for the powersats and their energy.

In any case, there would be need for central management and responsibility in so difficult a project, or else there would be an endless round of buck-passing. Centralization of this kind is much easier if only one nation takes on the whole project. There have been several large international technical undertakings, but many of them have come to grief. A good case in point is Europa, a multi-nation rocket program. Europa was designed in the late 1960s using technology mostly from the fifties. Britain was to build the first stage, France the second, Germany the third. These nations so disagreed with each other that the rocket never got built.

So a space community will not be a place for avant-garde or experimental approaches; instead, it will likely be set up along familiar lines accepted by all. In its technical development, it will rest on well-proven rockets and established techniques in metallurgy. Similarly, its society may well be founded upon such well-known ideas as the nuclear family, the privately owned home, and the use of money as a medium of exchange. There is much to argue that it will be a bastion of American middle-class values, excepting such changes as grow out of the colonists' **181** own experiences. This conclusion might not be romantic; but in this prosaic world, few things really are.

Can we speculate even further? Can we point to some Earthside community that is enough like a space colony to give us some idea of what may come? To answer this, we need only recall that early in this century, the U.S. embarked on a venture that in many ways will bear comparison to a powersat program. Here too there was need for the most advanced techniques, the best organization for the energies and vigorous determination of the whole nation. Here too our skills were transported to a remote and hostile land, pitted against severe difficulties of nature, focused within a tiny geographical area. The project was the Panama Canal, and the society that grew out of it was the Canal Zone.

The legend of Panama has long been part of our national lore: For centuries men had dreamed of a passage between the oceans. The French had tried, but were defeated by the immensity of the task and by the diseases of the jungle. Then the Americans came in. They cleaned out the yellow fever and malaria, brought in better machines, and proceeded to make the dirt fly. Where the French had failed, the Yankees succeeded.

And when the great effort had reached its successful completion, when the tide of activity and resolution had receded, there was the canal—and the Canal Zone. After the canal opened in 1914, there was little scope for ingenuity, few positions calling for exceptional talent and genius. The Canal Zone soon settled into a routine, almost somnolent existence marked by an awed reverence for the great accomplishments of the past together with an ongoing present embodying all the inspirational qualities of a center for government bureaucracy.

The Canal Zone was not long in acquiring all the amenities that advocates of space colonization would wish for their orbiting communities: spacious private dwellings, neatly laid-out towns and villages, schools, churches, TV and radio, restaurants, and shopping. As a counterpart to the Personnel Orbit Transfer Vehicles, there was a steamship line to the States; and like a space colony, the Canal Zone came to rely on this government-owned transportation for most of the amenities of life. Though milk and beer were produced and bottled locally and native fruits were eaten, much of the food and virtually all the clothing and household goods were imported. There were beaches, golf courses, even a community college—amenities that any space colonist might envy. The Canal Zone was not long in achieving the goal of the last chapter, a stable work force. Indeed, many Zonians came to live their entire lives there, and their children and their childrens' children as well.

For all that, since its inception the Canal Zone has been run as a company town. Few Pennsylvania coal barons could exert greater control over the lives of their employees. While this control has rarely been exercised capriciously, neither has it been particularly subtle or discreet. As Canal Zone Governor John Seybold said in 1956, "If you don't like it here, there's a boat leaving every Tuesday."

Salaries are high; that is one of the principal attractions. Canal employees receive 15 percent more than they would be paid in equivalent jobs within the federal government. At the same time, rents are low: $169 for a three-bedroom house. But no canal employee may buy land or own his own home. If fired, he and his family will have to move out and leave the Zone entirely within thirty days. Even while living in his home, he will find restrictions many people would regard as unacceptable. If he wishes to paint the walls in a color of his choosing, he will

Life in Panama during canal construction days. Top, *the docks at Cristobal*. Bottom, *the town of Tabernilla*. (*Courtesy Panama Co.*)

need government permission. If he invites friends to stay as house guests, he will need permission to have them stay overnight.

Every company town must have a company store, and the Canal Zone is no exception. Government commissaries sell food, clothing, shoes, furniture, and appliances at U.S. prices. The right to shop there is a privilege reserved for canal employees and their families and is jealously guarded. But these stores have a monopoly in the Zone, and it is government bureaucrats who pick the brands and the fashions. If the result is not nearly as dreary as a Moscow department store, neither is a commissary quite like a suburban shopping mall. The dissatisfied customer can shop in Panama proper, but the selections are usually even less varied and the prices much higher.

The company-town analogy is apt, for by law no Canal Zone resident can go into business for himself. Even the doctors and attorneys are civil service employees, and only the U.S. government can offer employment. This monopoly aids greatly in maintaining political control. Freedom of the press is not prohibited; the Zonians, after all, are U.S. citizens with constitutional rights. But free enterprise is, so there is the same result. The only sources of information are the government-run newspapers and other media. To be sure, there is a form of free press there, and it is legal: a local newsletter. The only problem is its blatant unprofessionalism, because none of the people involved with it are professional journalists. There simply aren't that many around and those that are, are hired by the government for its own publications. Is there freedom of the press? Technically, yes. Access to local, independent, professionally reported news? No.

In a political sense, the Canal Zone is unique because thousands of U.S. citizens live there without the ability to exercise their constitutional rights. This is not to say that they live under martial law. It is just that the U.S. has seen no crying need to let the Zonians vote for anyone locally. They have the right to vote, just no one to vote for.

They can register to vote, but only by absentee ballot and only in the stateside locale from which they came. A population of their size would have considerable clout in a single congressional district, but what the Zonians have instead is a gerrymander in reverse. Instead of living within district boundaries drawn to consolidate a political stronghold, they vote separately in hundreds of congressional districts throughout the U.S.

Real power lies wholly in the hands of the governor, who is appointed from Washington, and in his advisers. The Canal Zone residents do elect local councils, whose leaders can meet with the governor and offer advice. But these civic councils are not agencies of decision, for they have no authority. They cannot levy local taxes for needed civic improvements, for as with everything else, all spending power lies with the governor.

One well may wonder how such a system could exist today in the late 1970s. The answer is peculiarly relevant to the future of any powersat program, for the reason is: It was set up that way from the beginning. Indeed, at the start there even were features that are absent today, such as free housing and medical care. When the U.S. government determined that it would build the canal and would bring in the work force that would accomplish the task, it was but a short step to deciding that it would take on full responsibility for providing for their needs and even for their creature comforts.

The resulting system was deliberately set up to avoid having the character of a string of transient work camps. Instead, the emphasis was on providing stability for the skilled employees. Early on, the

housing

Terraced apartments in a Stanford torus. Will they be owned by a central bureaucracy and rented to the tenant-employees as in a company town? (Courtesy Ames Research Center)

government determined that the best way to do this was to set up communities to which married men might bring their wives and which would encourage bachelors to get married. If the resulting society was to be structured and paternalistic, this would be not a bad thing but a good thing. For a family to live in Panama would be adventure enough, let alone that they should suffer hardships or even inconveniences. The emphasis was on encouraging normal family life. If this meant that the atmosphere would be aggressively wholesome, that the main diversions were to be found at the YMCA, that the saloons and whorehouses were carefully tucked away outside Canal Zone boundaries—then so it would be. Stability and family life were the things to be sought, and these policies worked. To describe these employees, in the words of David McCullough (*The Path between the Seas*),

> Panama seems to have been the experience of a lifetime, almost without exception. The work, the way of life, the sense of being part of a creative undertaking so much larger than oneself, were like nothing they had known, as they openly and cheerfully expressed then and for the rest of their lives. . . . The

work was everything. Pride and joy in the work constituted the magic bond which held the canal colony together. There was no one who was not associated with the work. No one could live within the Zone unless he or she was a worker on the canal or a member of a worker's family. The entire social order existed solely for the work and it rewarded its members according to their importance to the work.

Even in those days, while the canal was under construction, there were some who were disturbed at what they saw. To the American diplomat William Sands, this new civilization was "a drearily efficient state, a mechanization of human society. . . . From a railway car one could tell by the type of furniture and the color of the hammock swings the salaries and social standings of the occupants of all the houses." Others expressed concern over what looked to be a successful experiment in socialism. After all, it was the government that owned the railroad and telephone lines, the housing and the commissaries, even a hotel.

This view was challenged by a Socialist Party member, a mechanic who had been with the canal project almost from the start: "First of all, there ain't any democracy down here. It's a bureaucracy that's got Russia backed off the map. . . . Government ownership don't mean anything to us working men unless we own the Government. We don't here."

I have dwelled at length on the Canal Zone because it may well indeed be the model for a space community. With the requirements of the power satellite and of lunar resources being no less demanding than those of Panama in an earlier day, the evolving space community could well develop in a very similar direction. It would be run by an all-encompassing government bureaucracy, which concentrates all power into its central administration, while conferring benefits on its employees. The tasks of powersat construction could become as standardized and routine as those of maintaining and operating the Panama Canal. In such a space society there would be no unemployment, no poverty or economic hardship. But neither would there be challenge or change. Like the Zonians, the space colonists would cherish the memories of the great feats of the past, while living in a present that would be, ironically, mundane.

The very success of the Canal Zone's government shows how well such a system can work. Moreover, the Canal Zone government would be well suited to the size of a space community. At the peak of canal construction, in 1913, there were 5,362 skilled employees, whose latter-day counterparts might live in space. Today the number of canal employees stands at around 3,500. Such a work force would suffice to build two to three powersats per year from lunar materials, an eminently reasonable rate.

Withal, the true situation will doubtless prove nowhere near so bleak. As activity in space expands, as more and more people go there to live, it will become increasingly possible for entrepreneurs to set up small companies, which will serve the needs of the colonists. The growth of space settlements may make it increasingly difficult for the government to maintain a Canal Zone-type control there.

The reason will be the colonists' continuing need for the comforts and amenities of life, ranging from toothpaste to appliances. In the Canal Zone, this need merely furnishes cargo for the government's shipping line. In space, such imports will call for rocket transport, which likely will always be several times more costly than air freight. With a growing space population, in time it will become possible for local entrepreneurs to compete with these imports by producing them locally.

It might be that the government would challenge such activities and seek to limit them. However, there would be no shortage of Earthside individuals and groups willing to challenge the government on this issue in court. Again, the government might try to pre-empt these opportunities by setting up its own subsidized production centers in space. But that would not only involve the government in a myriad of small businesses of a type it would be ill-equipped to run, it also would fly in the face of long-standing NASA policies, which seek to encourage private investments in space activities.

In a static space society performing well-defined tasks, the Canal Zone analogy would be obvious, but this is not necessarily true for a continually growing human presence in space. With the increasing needs of such a community, the government might welcome the opportunity to turn matters over to private enterprise, rather than absorb the high subsidies for rocket transport. There thus could come to be new meaning in our association of liberty with ''the rockets' red glare.''

Even with lunar resources, the power satellite alone may indeed need no more than an orbiting Canal Zone, with ten thousand people living in space. After the initial buildup, the task would be quite stereotyped—to build so many powersats per year. But the powersat is not the only large-scale activity that can draw numbers of people to space, merely the one that is perhaps most near-term. There are others as well, and in their eventual growth may lie the hope for human diversity in space.

The advent of powersats should thus indeed bring space colonies. But they will not be the colonies of which space advocates dream, for they will be strongly molded by their character as centers for the powersats' people. To fulfill these dreams, space colonies must attract large numbers of Earth people, not because a project needs workers, but simply because space colonies have become enticing places to live. The orbiting powersat colony can serve as the architectural model for these later, larger colonies. There are activities which will lead to millions of colonists—not a mere ten thousand—and the most important of these could well be space industries and space tourism.

Chapter 10

Lovers, Colonists, and Explorers

Space tourism might begin even before there is a space colony, with the building of an orbiting vacation resort. In 1967, at a conference of the American Astronautical Society, the hotel entrepreneur Barron Hilton (son of Conrad) stated that if space transport costs fell to $5 per pound, he would build a hotel in orbit. Actually, inflation has turned Hilton's $5 per pound into what in today's dollars would be more like $10. With reasonable provision for baggage and for the food and oxygen to be used by the orbiting tourists, a round-trip ticket then might cost $4,000 or so. This is not much more than twice the cost of a round-trip transatlantic ticket on the Concorde and is similar to the costs charged for many cruises by ship. Since the Cunard Line has successfully sold tickets for their most luxurious round-the-world cruises at up to $97,000, the potential is obvious.

The possibility of a space hotel then rests, as does so much else in space, on the availability of low-cost rockets. The scramjet, discussed in Chapter 5, is an entirely new form of engine, which offers the prospect of an aircraft that will fly to orbit as if it were a fast jet plane. The day will come when vacationers can reserve seats on such a craft on a flight out of Miami to orbit.

What a flight that would be! Using turbojet engines, the plane would fly to open ocean, where the sonic boom of supersonic flight would not be annoying. At Mach 2 the jets would shut down and the scramjets take over. Passengers would not notice much, but they would continue to feel the acceleration as the plane gathered speed. Through their windows they could see the sky turn a deeper, deeper blue. Soon it would shade off into virtual blackness, and the curvature of the Earth would be evident. There would be no vibration, no harshness of ride, but on the cabin bulkhead a digital Machmeter would display the increasing speed: Mach 6, Mach 8 and on up as high as Mach 14.

At that speed, and at 120,000 feet altitude, it would be the turn of the scramjets to shut down. In the rear of the aircraft, rocket engines now would thunder to life. With the hardest part of their job already done, the rockets would soon drive the travelers the rest of the way to orbit. The craft would dock with the orbiting hotel, and the passengers could deplane and seek their staterooms.

In the hotel lounge would be large windows offering excellent views of both Earth and space. With telescopes and cameras, or merely sitting in comfortable chairs, the vacationers could sit enthralled as their planet passed below. Awesome would be the sunrises and sunsets, every ninety minutes, as the Sun spread its waxing and waning light across the world or made incarnadine the lower air. The land would offer a never-constant panorama: now the deep reds and yellows of the Sahara, then the dark green of the Amazon rain forest, and again the cloud-speckled blue of the sea. The Himalayas—those majestic mountains spanning the view, yet which one is the Everest which Hillary and Tenzing struggled to climb? The astonishing crystal greens of shallow tropical seas near Bermuda, near the islands of the Pacific. An anvil-shaped thunderhead towering above an expanse of white cloud over Kansas. The great cities of America and Asia, glowing with light amid the nighttime darkness. The feathery watershed of a great river in spring. And always, on the horizon, the light blue where the sky is as seen from below.

When they turn away from their Earth-watching, the tourists will find a number of unique attractions. Since artificial gravity can be set to any level simply by controlling the rotation rate, many Earthside sports and games will take on an entirely new character. For instance, there could be a circular jogging track, on which people could run in the direction opposite to that in which the track is spinning. As a jogger would speed up, his weight would go down, since he would be counteracting some of the centrifugal force that was holding him down. If his pace was fast enough, his weight would vanish entirely, and he might find himself kicking vigorously while slowing rising into midair.

Nor would humans be the only ones to enjoy this; dogs could too. Fans of greyhound racing could find much to cheer in an orbiting version of Hialeah, where reduced gravity would greatly

A spaceliner approaches the hotel in low orbit. (Courtesy Don Dixon)

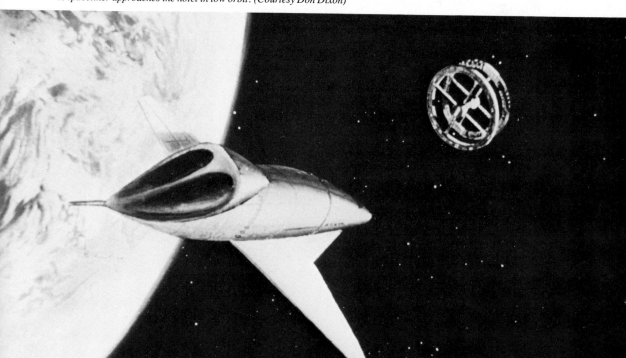

View of the orbiting hotel. (Courtesy Don Dixon)

lengthen the graceful strides of the competitors. In fact, an entire orbiting menagerie might develop. What about the idea of turtle races? Think what they could do without the weight of their heavy shells. How would kangaroos fare, or monkeys swinging from the branches of trees? Or a new version of Mark Twain's "celebrated jumping frog of Calaveras County"?*

Water sports would take on an entirely new character. A low-gravity swimming pool would resemble an enormous slowly rotating barrel with water all along the inside of its periphery. Someone swimming there could look up and wave to his friends directly overhead, for he would see the water arching uphill to left and right, then continuing overhead to form the interior of the barrel. Swimming with the aid of flippers need not be limited to the water. A swimmer could launch himself upward from underneath, break the surface like a dolphin, and then continue upward into the air. The flippers then could serve as little wings, further propelling and steering the swimmer as he sought to reach the center of the barrel. Ordinarily he would not succeed, and would slowly lose speed, then fall back. But a few would achieve zero-g. There they could float and relax, watching the scene around them.

Human-powered flight could be a most novel and pleasurable sport if the orbiting resort grows large enough. The aircraft would resemble hang gliders, but with a propeller turned by bicycle pedals. The daring aviators could start their flights near the center of a large rotating enclosure, where the artificial gravity or centrifugal force would be weak and the pilots could get a good start. Thereafter, the challenge would be, "how low can you go." By swooping low, outward from the center, the artificial gravity would increase, and pilots would need more and more effort to maintain the margin of power for a return to the central, low-gravity regions. Of course there would be no danger if

*Every July there is a jumping frog competition in Calaveras County, California. The winning jumps have been around twenty-one feet.

View of Earth from the hotel lounge. Britain, Ireland, the Channel, Denmark, and France all are visible. (Courtesy Don Dixon)

Low-gravity swimming pool resembles the inside of a barrel. People can swim not only in the water but also in midair with the aid of paddles or small wings. (Courtesy Don Dixon)

someone dipped too low; he would simply glide to a landing on the inside of the enclosure. Still, this would prove quite embarrassing on occasion, especially to someone who wanted to show off.

Even with less elaborate wings, tourists will find opportunities to take advantage of zero-g so as to fly after the fashion of birds. Many vacationers will practice for hours or days till they understand how to control their flight, and thereafter will treasure the home movies or Polaroid photos that will attest to their skill. Yet there will be those for whom these memories will be far from the strongest. Some will seek activities of a very private nature, which may leave them little time for the views from the lounge or for sports.

Among the most popular attractions will be the opportunity for sex in zero-g. Here will be the chance for lovers to try out all the positions that proved uncomfortable or difficult back home, even in a waterbed. The Hindu love manual, the *Kama Sutra,* will doubtless be popular reading; with good reason, it will be on sale in the airport departure area. Some *Kama Sutra* experts like to try out difficult positions in the water first, but they will find that space is much better.

Neither partner can be on top or on the bottom when there is no up or down. It will take a lot of cooperation for novice couples to stay locked together, and this will prove easier with the aid of an elastic support. This will be a situation unimagined in the Middle Ages, when the lady of a knight wore a chastity belt; what will be needed here is an unchastity belt, to keep the couple coupled. It may be of pink latex or nylon, with holes for four legs, somewhat resembling a set of underpants to be worn by a baby elephant. However, it will be worn by a rather different animal, what Shakespeare referred to as the beast with two backs.

More experienced couples will dispense with such artifices and rely on the woman to wrap her legs around her man, then to move her body in appropriate ways. A man may prove quite inventive, grappling his woman's thighs with his knees so as to thrust into her. Others will find various ways for a woman to sit astride her man, facing either forward or backward. And so the long night will pass.

And how good it will be, after the loving, to fall asleep still holding each other in one another's arms, with no gravity to press the weight of one heavily on the other, or to cut off circulation in arms that would embrace.

As time goes by, there will be many couples who will make love in this way, and many thousands of people who will sample these pleasures of life in space. In the meantime, the space colony will be developing and its people building their powersats. There will naturally be the question of bringing to the colony the pleasures and amenities of such an orbiting resort. With an imaginative colony leadership, they could be the basis for the next phase of colony growth.

As the program for building powersats expands, the need for new colonies will expand apace, and a solid understanding of how to provide for a space community will develop. There will be a large corporation in space, or perhaps several such corporations, which may actually be departments of Earthside governments. These corporations will run the space transport and the centers for space construction. In addition, their orbiting ''Panama Canal Zone'' colonies will provide homes for the people who live in space; and the corporate management will have reduced to standard practice the means for supplying these people with their needs, without recourse to large-scale transport from Earth. When a new space community is to be built or a section of agricultural land expanded, the board of directors can vote to have these things done under fixed-cost contract. It will be the same as the routine, standard way an airline orders new jet planes.

We may imagine that the corporation that produces and leases the powersats operates a very profitable business on a steady, even financial keel. It has ample flow of cash, and a secure, constant **193**

Sports in zero-g. (Courtesy Ames Research Center)

income from its powersat operations. It is then that the corporation would seek to grow and expand by moving into related areas of endeavor. One alternative would be the real estate business.

There is no reason why real estate development in space should be all that different from its more traditional Earthside counterpart. On Earth, areas of land acquire value when people want them and are willing to bid for them, as at an auction. In space colonies, real estate will have the intrinsic value associated with the capital costs and the energy needed to produce it, but there can be as much of it as one wishes. Land in space colonies will be like autos or refrigerators, which are sold at fixed prices and produced to meet the demand.

The simplest act of space colonization would call for no new construction at all. It would simply involve a change of policy: Instead of the corporation owning the people's homes and renting them to its employees, the tenants would be granted the opportunity to acquire ownership and a clear title. For those who had been living in space the longest, their rent payments of past decades might simply be counted for bookkeeping purposes as mortgage payments; they could own their homes free and clear. Even the most recent newcomers could have their rental contracts converted into mortgages, and rental payments of past years could be counted as mortgage payments.

Even so simple a change would mark the end of an era. No more would the space-dwellers be regarded as wards of an all-powerful government. No more would there be a central administration to take responsibility for all phases of the people's lives. It would mark the end of space communities as company towns, as extensions of bureaucratic fiefdoms. It would mark the beginning of a true space economy, aimed not at serving the needs of Earth but rather at serving the people who live in space.

The end of universal government ownership would also mean the end of the policy of tight control over who could and could not live in space. No more would there be careful selection of applicants to fill those jobs and only those jobs that the government deemed worth providing. No more would an employee have to leave the colony if he quit his job or was fired. Instead, there would be opportunities for people to make a living by private employment, providing new goods or services. An Earthsider, long attracted to space, could arrange to buy the home of a retiring space colonist and move there. With a loan from the Bank of the Colony, such a newcomer could work to set up his business, hoping his fortunes would prosper. Home remodelers, builders, general contractors would all find themselves newly welcome in space.

There would be need for caution here. A too-rapid changeover to private ownership could spark a speculative real estate boom. One can well imagine that the mid-twenty-first century will see numerous millionaire space enthusiasts who have whetted their appetites with visits to the space hotel but whose real wish is to live there permanently. In a few days or weeks they might bid the prices of space homes up into the millions as they competed for the few thousand or so newly available properties. Many a colonist would find himself rich overnight, at least on paper, as his modest $50,000 home or condominium happened to catch the fancy of some wealthy Earthsider. By selling, he would realize a quick financial killing; but the space activities would be deprived of his skills as he left for Earth with his money. In the span of a few months the space colony could change its character completely, from a home for valued employees to a center for frenzied real estate speculation. The eventual bursting of this financial bubble then would prove most unpleasant to all concerned.

So it would be important that from the start there be an adequate supply of real estate in space colonies. This would mean building more such colonies, whose land and homesites would be made available to Earthside buyers. There would be other land, too, for light industry and for shopping malls. The day of the company store or commissary would be at an end.

Large space colonies may be miles in extent.(Courtesy Don Davis)

As a minimum, such ''for-sale'' colonies would be built as a pressure shell of strong metal, with mirrors and window areas to let in sunlight. They would have coverings of lunar soil for radiation shielding, interior atmospheres, more soil on the inside. They would also have equipment for circulating electricity, air, and water while keeping the latter two fit for human use. The interior layouts of such colonies would vary. For those intended as farmland, there would be little more than flat soil. At the other extreme luxurious colonies for those with expensive tastes (and wallets) would offer lakes and streams, marinas, golf courses, elaborate recreational areas, and even equestrian trails. Most likely the land would be sold in a condition ready for development, but without pre-existing buildings or other structures. It would be the responsibility of the buyers, whether as private individuals or commercial firms, to hire contractors to put up their homes and businesses. The colony government would still have its hand in this, but it no longer would be an all-controlling power. Instead, it would be like a real estate developer who seeks to encourage buyers who will contribute to the growth of the community.

Apart from the glamour of living in space and being a part of humanity's reach outward, there would be an eminently practical reason for Earthsiders (especially those from Chicago or the Northeast) to prefer life in a space colony. A colony could easily be made to have the best climate available, milder than Hawaii or Southern California, entirely free from storms or other surprises, never ceasing to be warm or bright. The weather, of course, would be fully controlled. Nor need the scenery appear unnatural, artificial. If demand booms, it will be quite possible to build colonies that are miles in extent, with a blue sky and clouds, and with room for a million or more. If so few as a million Earthsiders were to take out mortgages averaging $100,000—the price of a home and lot today in many desirable parts of Southern California—that would be $100 billion. That would provide for quite a bit of space construction.

In 1975, the artist Don Davis prepared a set of color paintings showing visions of the interiors of the largest possible space colonies. He showed parklands and green forests, rivers and hilly uplands, even a bay full of small boats and spanned by a large suspension bridge. Such ideas may appear as mere fantasy, but they are no such thing. They will be entirely necessary if space colonies are to build a reputation for quality and desirability.

Perhaps the activity most resembling this kind of space colonization today is the building of new communities in the deserts of the Southwest. Several such developments have been built by a subsidiary of McCulloch Oil Corporation, which like our fictional powersat firm, is a successful energy corporation that has diversified into real estate. The subsidiary, McCulloch Properties, has built such towns as Lake Havasu City and Fountain Hills, Arizona; Spring Creek, Nevada; and Silver Lakes, California. Over the wasted deserts where twenty-mule teams once hauled borax from Death Valley, there now rise these attractive and much-sought new communities. Lake Havasu City is

Interior of a large space colony. (Courtesy Don Davis)

particularly well known; it is there that McCulloch rebuilt London Bridge, having shipped it from England. It is enough to make one wonder; surely the Golden Gate Bridge cannot forever meet the needs of San Francisco. Will it one day be dismantled and rebuilt in a future Lake Havasu City?

If a space community follows the pattern of the McCulloch developments, its amenities will be lavish indeed. For starters, there will be lakes well stocked with trout and coho salmon. Around the lakes will be boat docks and launching ramps, sandy beaches, barbecue pits, and grassy park areas. Many families will enjoy access to the lakes from their own backyards. Because the landscaping can be in whatever way the developers want, it will be easy to design a lakefront community as a number of long, low, sandy peninsulas extending into the lake. Each peninsula will be wide enough for a road down its length, with lots on either side backing against the water. The sight of brightly colored sailing craft, or of catamarans with aluminum hulls, will be just part of the local scenery.

There will be plenty for the sports-minded. It will not be hard to bring in championship tennis and golf pros, to manage these sports activities, and to arrange special events or tournaments. A professionally designed and landscaped series of golf courses will permit 9-, 18-, or 27-hole play, regular or championship. The ready control over landscaping will mean a challenging layout for every shot. For those who are not so proficient, a large putting green and driving range will permit people to sharpen their game. The game of golf will actually be more challenging in a space colony because the gravity will be artificial. Since it will be provided by rotating the colony, there will be an effect due to the rotation known as the Coriolis force. It will cause golf balls to slice to left or right even if hit straight down the fairway, and even the most proficient of golfers will virtually have to relearn the game to correct for this.

Every golf course must have a clubhouse, and this one will be no exception. Perhaps it will be a two-story structure on a rise of land, with excellent views of the colony interior. There will be a cocktail lounge and an excellent restaurant, a main lounge with high cathedral ceiling and lavish floor space, a fully equipped recreation room with gymnasium facilities, locker rooms, bowling, saunas and massage rooms, whirlpool baths, as well as meeting rooms and movie theaters. Just outside the clubhouse will be the basketball and handball courts, as well as the tennis, and the swimming pools. An Olympic pool with three levels of diving, a Jacuzzi, a small pool for children, all with wide concrete decks and ample beach chairs, with a snack bar, will complete everyone's enjoyment.

Elsewhere will be a hotel with convention facilities, riding trails and stables for horses, and bicycle paths. These may be in a separate equestrian park. And all these facilities can be provided with no users' fees charged to the community residents or to their guests.

When building on a lot, the homeowner will have a choice of architectural designs, which permit easy construction with the aid of robots. A builder may control a team of robots as though he were pharaoh driving a gang of slaves in a Cecil B. DeMille movie. This method will likely mean one of a couple dozen standard home layouts, available in different colors or styles of exterior trim, built from factory-manufactured modules. Custom-built homes will cost more. Construction will be straightforward and building codes simple, for there will be no storms or ice, no natural hazards, not even groundwater to leak through a foundation.

Yet if these colonies are to prosper, it will not be enough to establish them as resorts or real estate developments. Their success will call for much more than wealthy or independent space buffs buying homes there for reason of their novelty. Indeed, it would prove a very bad mistake to promote space colonies simply for reasons of fashion, for fashions change all too quickly. Biarritz may have been *três chic* last year, but this year no one would be caught dead there, dahling.

rmland in a large colony. (Courtesy Ames Research Center)

A new real estate development in a colony, built with the aid of robots. (Courtesy Don Dixon)

Fortunately, there appear to be at least two industries that can supply the economic underpinnings for these orbiting cities. One of these would be the design and building of spacecraft. The powersat will be only the first of these. Just as the settlers of New England took advantage of their forests to develop Boston and New Bedford as important centers for shipbuilding, the space colonies will be advantageously placed to build their own ships.

By developing skills as instrument makers, propulsion specialists, electronics designers, the space colonies increasingly will be able to compete with the Earthside centers for spacecraft development. Among the people who will go there to live may be many astronomers, and other scientists as well. Great observatories and physics laboratories can grow there to take advantage of the vacuum of space and the unimpeded views of the stars.

The second major industry would be information. In the next century many people will spend much of their working days at computer consoles or telecommunications terminals and will be free to live anywhere they please. The use of communications satellites will allow them to do their work and keep in touch with colleagues and associates without having to leave their offices. If such people are attracted to life in the space colonies, it may be less for the golf courses or even the zero-g sex as for the superb, unequaled communications facilities they may command. Consultants, writers, analysts, financial managers—these may be only a few of the people, many of them self-employed, who will be able to make their livings in the space communities.

These communities can help facilitate the classic dream of astronauts setting off for distant planets on missions of exploration. This is a dream inspired by the historic voyages of Columbus, Drake, Hudson, Cook. It expresses the hope that as the ship with sails set was the very symbol of the Age of Discovery, so might the rocket, outward bound, serve as the harbinger of a new Renaissance.

The present era and the decades ahead are proving indeed to be a great age of space exploration, but the explorers are instruments and robots, not men. The people will follow in their wake, but initially there will be little of the grand romance of an H.M.S. *Endeavour* or *Golden Hinde*. Space exploration has much more in common with exploring the Antarctic than with the voyages of Columbus, and one does not venture to Antarctica in crude wooden caravels manned by illiterate sailors from the docks of Cadiz. But for good scientific or economic reasons, we will place large parties in McMurdo Sound or Prudhoe Bay. The same will be true when we send groups of people into the space beyond the Moon.

It was oil, ten billion barrels of it, which brought people to Prudhoe Bay. The space activities of the future will not seek this resource, but they will rely on what to them will be as valuable: water, carbon, nitrogen. With these it will be possible to go beyond the building of structures merely of metal and silicon. There will be new structural materials, stronger and with superior properties. There will be the opportunity for space agriculture. Above all, a ready source of these prizes will point toward the end of dependence upon Earth for a space project's most vital needs. It will become possible to think in terms of a civilization in space.

The nearest source of water could well be the Moon. For nearly twenty years there has been a solid scientific basis for speculating that water may exist there. The Apollo missions found the Moon to be exceedingly dry, so much so that by comparison the harshest deserts of Earth would appear a rank and growth-choked swamp. Not only was there no evidence of water or ice; the rocks lacked telltale chemical signs of even having ever been exposed to water.

However, all lunar explorations to date have been carried out near the lunar equator, in regions which for billions of years have felt the blasting heat of the Sun. In 1961, a young Caltech scientist named Bruce Murray pointed out that near the lunar north and south poles are permanently shadowed regions where the Sun never shines. An example would be the interior of the crater Peary, whose high rocky walls block all direct sunlight. In such shadowlands there is only the feeble light of sunshine reflected from distant uplands. Nor are the shadow areas small in extent; they cover some one-half of a percent of the total lunar surface, which is about the size of the state of Michigan.

It was the proposal of Murray and two colleagues that these shadowlands could serve as "cold traps." Temperatures there would always be hundreds of degrees below zero. Any water on the Moon would freeze out and be trapped there. Over geologic eons, sizable quantities of water might accumulate—tens of billions of tons, according to the most recent estimates. This same Bruce Murray in 1976 became director of NASA's Jet Propulsion Laboratory, the leading center for lunar and planetary exploration. It is not surprising that he has advocated a mission known as Lunar Polar Orbiter, a lunar satellite which would pass over both poles. It would carry an instrument known as a gamma-ray spectrometer, which would detect water in the cold traps, if it exists.

Much of this presumed lunar water would have been brought in by impacts of meteorites or comets. A common class of meteoroids is a type known as carbonaceous chondrites; these typically have a water content of 3 percent. Much larger quantities of water are known to exist in comets; a single comet could readily have brought a billion tons of water to the Moon.

While lunar water is a matter for speculation, the water content of comets and meteoroids has been measured by observation. As an oil geologist would say, lunar water falls under "possible resources," but water from comets and meteoroids would be "proved resources." It thus is quite significant that recent advances in astronomy have made it possible to identify the sources of meteoroids and to find bodies in space that actually are immense carbonaceous chondrites. Moreover, it now appears that a rich source of materials lies close to Earth's orbit, in many cases scarcely more difficult to reach than the Moon. This resource would not only provide water, but carbon and its compounds (as the description "carbonaceous" attests) and quite probably nitrogen.

These are the Apollo and Amor asteroids. Most asteroids are small rocky bodies orbiting the Sun between Mars and Jupiter, but a few venture closer in. One of them, named Icarus, passes closer to the Sun than the planet Mercury. Those that cross the orbit of Earth are called Apollo asteroids*; those that pass just outside Earth's orbit are the Amor asteroids. Some twenty-eight Apollos have been

---

*The name has nothing to do with the Apollo program. The first such "Earthcrossing asteroid" was discovered in 1932 and named Apollo; it gave its name to the group.

Location of meteoroid impact scars in Canada. Filled circles are known impacts; open circles are probable ones. (After Richard Grieve; map by Gayle Westrate, California Institute of Technology)

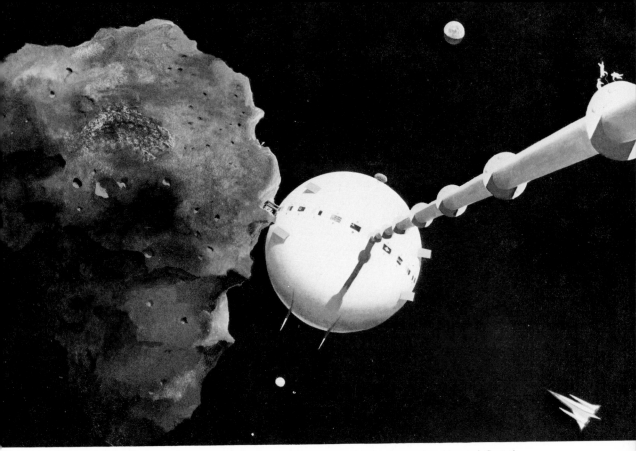

Approaching an asteroid with a mass-driver engine. (Art by Chesley Bonestell; courtesy Ames Research Center)

discovered, and more are being found at the rate of at least four per year. Various astronomers have estimated their total number, and the estimates tend to agree: There are some five hundred to one thousand such bodies larger than a kilometer in diameter.

In 1937 one such body (Hermes) passed within a half-million miles of Earth, a very near-miss on the cosmic scale. Had it hit, it would have struck with the energy of ten thousand large hydrogen bombs, blasting a crater some twenty miles in diameter. Indeed, such impacts occur three or four times each million years. Most of them are in the ocean, but land impacts are far from unknown. Arizona's Meteor Crater is an example; a much larger one is the forty-five-mile-wide Lake Manicouagan in northeastern Quebec. Other old impact scars are the Ries Kessel in Germany, the Vredefort Ring in South Africa, and Gosses Bluff in Australia. Canada has an especially well-preserved region of old impact craters. The scars have been preserved in the deep granite of the Canadian Shield and have been uncovered by the action of glaciers. A map of such impacts, prepared by Canadian geologists, makes their country look like nothing so much as a portion of the planet Mars.

These impacts show that there must be a continuing supply of new Apollo asteroids; the existing several hundred would all hit Earth (or Venus) in the course of a few hundred million years. They are, in fact, ancient comets.

Comets come from outside the Solar System, from a region of interstellar space known as the Oort Cloud. A passing star can disturb the motion of some of them, causing them to enter the Solar System. As they pass by Jupiter or Saturn they may be ''captured,'' pulled by gravity so as to become a permanent part of the Solar System. A very few experience close encounters with the gravity fields of Mars or Earth. Then their orbits can change so that they become Apollo asteroids.

Just this type of process seems to be taking place with an object known as Comet Encke. Its orbit is intermediate between that of an ordinary comet and that of an Apollo asteroid. What's more, in recent decades it has shown definite signs of weakening in activity. A comet's brilliant display results from its content of easily vaporized gases. These gases are frozen in the cold of interstellar space but vaporize when a comet nears the Sun. When they are all gone, a comet ceases to be active and becomes instead a type of asteroid. When Comet Encke becomes inactive, after perhaps a hundred more orbits, it will be an Apollo asteroid and may one day strike Earth. It is unsettling to realize that events in interstellar space could lead to a catastrophe that might wipe out the New York metropolitan area.

Can asteroids be retrieved, transported to Earth orbit to be mined for their valuable water and carbon? In 1964 Dandridge Cole and D.W. Cox wrote a prophetic book, *Islands in Space*. They proposed that asteroids be used as resource mines, hollowed out or otherwise processed in order to build space colonies. Indeed, they freely used the terms ''space colonies'' and ''space colonization.'' Their preferred transport technique was the mass-driver. As a means for launching lunar payloads, the mass-driver was already old; it had been proposed by Arthur C. Clarke about 1950. What Cole and Cox proposed was that it serve as a new type of rocket, or reaction engine. Fitted to an asteroid, it would eject chunks of matter taken from the asteroid itself. The asteroid thus would slowly be consumed as it would wend its way earthward.

Much valuable material would be lost this way, so it is fortunate that there is another way to proceed. This method would use large banks of ion-electric thrusters, such as will move powersats from their construction sites to geosynchronous orbit. An asteroid could be fitted with a chemical plant that would extract oxygen from the minerals in which it is chemically bound. The oxygen, and only the oxygen, would serve to run the reaction engines. The valuable hydrogen and carbon compounds, as well as refined metals, would be saved and not ejected.

Because ion-electric engines give superior performance, this approach would offer several advantages. Much less asteroidal material would be lost as rocket exhaust. The exhaust itself would be a gas, not a hazardous stream of pellets or slugs of rock. That material which is ejected would be the most common of space resources, oxygen. Best of all, the asteroid would be chemically processed into valuable goods even while it was flying en route to Earth's vicinity.

In an era of concern over resource limits, it is comforting to know that the potential of the Apollos and Amors is not small. If these are regarded as fisheries whose stock is continually replenished, then on the average there are 400,000 tons of new Apollo/Amor asteroids introduced per year. If the Apollos/Amors are seen as a resource to be mined, they could supply a million tons per year for something like ten million years. It should be a long time before there is an Apollo/Amor shortage.

With the tapping of resources from the asteroids, the last requirement will be met for a long-term growth of the space colonies or communities. When these communities are built and settled, the human presence in space will be well on the way to expanding beyond the early beachheads won by the builders of power satellites. Yet in a sense this phenomenon will not be new, but something quite old.

Exploring an asteroid. (Courtesy Don Dixon)

Once maneuvered into Earth orbit, an asteroid could be mined for its resources. (Courtesy Don Dixon)

The first settlers in Virginia came to grow tobacco to send to the mother country; how much of the production of today's America finds its way to England? San Francisco was founded as a seaport, a center for shipping; yet today in the Bay Area one of the most important industries is electronics, and many people there have never set foot on a ship.

The space communities will grow for the same reason other civilizations have grown. They will have abundant sunlight and solar energy, rich resources from the Moon and asteroids, talented and creative minds to seek new opportunities. Like pioneers settling a fertile and well-watered valley, the settlers in space will be able to see the fruits of their efforts. Here will be no outcasts, sent to work a stony soil and to wrest a living from unfavorable lands. Here instead will be youthful, energetic people, confident in the assurance that their efforts will prosper.

The space immigrants will come from many nations. When the colonization of space is thrown open to all, there will be no restriction as to nationalities. It will once again be the opening of the American West, with diverse peoples coming to win a new land. Even at the early stages of the enterprise, there will be need for a colony government. As the population grows, and as the space economy becomes more and more self-sufficient, the human presence in space will become less and less that of a colony, a mere appendage of the mother planet. Increasingly the space-dwellers' government will assert itself in the councils of the world, speak for its own interests. Slowly, gradually, the space colony will evolve into a nation.

It will be a new type of nation. The newly arrived residents will not be like settlers on the western frontier; rather, they will be like homeowners moving into a new suburban subdivision. Nor will residence in space long remain a matter of golf courses and sports facilities. As with new communities in the desert, these amenities will give the space enterprise a reputation for quality, but there will then be the opportunity to construct much less lavishly appointed developments offering lower price tags to buyers. And this too will not be new. The first autos were expensive, affordable only by the wealthy, but they provided manufacturers with the experience which soon allowed Henry Ford to sell the Model T to Everyman. Airline tickets were for many years quite costly, but once the airlines had grown sufficiently, they were able to offer discount fares for all.

It is in this fashion that the human presence in space will grow. Yet that will not be the end of the story, but merely a new beginning. The influence on human thought will be profound. As the prospects of America embodied all the hope and optimism of nineteenth-century thought, so may the new and vastly larger frontier create new hope in the next century.

The people of space will not be demigods, nor will they be sybarites and hedonists. Rather, theirs will be no more than the strengths and weaknesses of any other people. Their material prosperity will be remarkable, but in the long run they may be cherished more for the hope they will give humanity: The hope of new horizons, of new opportunities, of a better life in a new land. All this will be real, it will be possible, it will exist—it will be theirs.

And the frontier will stretch ever outward. Inevitably, the colonists' thoughts will be drawn to the stars. The space-dwellers will not be left in ease and luxury, for to them will fall the great but unaddressed human challenge. It will be they who will grapple with the unanswered question: In the vast Milky Way galaxy, in the cosmos of which it is a part, what is the significance of man?

This is the challenge of the stars, and this is the question that will lie before them.

CHAPTER 11

The Fermi Paradox

It is part of the lore of science that one evening at a dinner party in the year 1943, the physicist Enrico Fermi startled his companions by looking around and saying, "Where are they?" When asked "Who?" he replied, "The extraterrestrials." His point was that if advanced interstellar civilizations really exist, they should be readily detectable, and their denizens might reasonably be expected to make occasional visits. Then, the absence of such visits or of evidence for such civilizations could be regarded as a surprise, a paradox.

Or, to put it another way: If life is common in space, and if intelligent life is far from unknown, then in the fourteen-billion-year history of the Milky Way, one might imagine that some culture has swept across the Galaxy in the grand manner of Alexander, Genghis Khan, the conquistadores. It then is not entirely obvious why our interstellar surroundings should appear to be in a wild state, as untouched as a wilderness preserve.

In Chapter 2 there was an estimate of how many of our Galaxy's stars may be abodes for advanced forms of life, including intelligent life. We saw that the star cannot have any kind of large binary companion, or the formation of planets will most effectively be prevented. The star must evolve and grow brighter at the right rate, neither too slow nor too fast; a life-bearing planet must be of the right size and form its atmosphere neither too quickly nor with too much delay. In addition, of course, the planet must be at the right distance. From this point of view, a planet will support life only if it and its star actually turn out to be a very close replica of the Earth and Sun, a replica that is to be achieved purely by chance, by the random workings of nature. As also stated in Chapter 2 it is likely this happened only for one in every 227,000 stars in the Galaxy, there then being some 880,000 "good stars" in all.

If we seek the homes of extraterrestrial civilizations, these stars are the candidates. In order to understand more clearly what we may expect of them, it is appropriate to discuss the difference between ''hang-ups'' and ''instabilities.''

As previously stated, a hang-up is an obstacle that effectively prevents or long delays the rise of an intelligent civilization. The failure of a star to form planets would be one; so would the failure of any planet to possess conditions suitable for life. The long span of eons required for the evolution of advanced brains represents another hang-up. Still another may be posed by geography. A planet may be completely water covered. It could develop high forms of life such as whales and dolphins, but these would not be candidates for the founders of technical civilizations. There can be no fire in the sea, and fire is the art that first distinguished man from the beasts. Less anthropomorphically, an ocean environment provides little novelty or challenging variety, which appears to give the strongest evolutionary advantage to high intelligence. On such planets, the rise of civilizations may have to await the rise of continents emerging from the sea.

An instability, by contrast, is something that hastens the rise of civilization. We may think of a rock perched atop a mountain; if it begins to move due to its own instability, it will rapidly gather speed and may soon sweep all before it. It is a characteristic of instabilities that they advance in a relatively short time. For instance, planets may not form at all (hang-up) but if they do, their formation will take no more than a few tens of millions of years (instability). On Earth it took four billion years for life to evolve to the point where there were complex multicellular organisms with skeletons or hard body parts, but it took only the relatively brief hundred million years or so of the Cambrian epoch for

An earthlike planet, of which there may be only one for every 227,000 stars. (Courtesy Don Dixon)

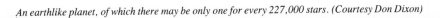

life to diversify and flourish into very nearly the variety we know today. This flourishing of life was due to another instability, which operated when the buildup of an ozone layer in the atmosphere allowed life to emerge from the protection of the sea.

The rise of intelligence and civilization, if at all possible, certainly must count as an instability. Ten million years ago the smartest animal alive was a species of East African or Indian ape known as *Ramapithecus;* today there is the glory of you and I. (It is an exercise for the reader whether this indeed constitutes progress.) One million years ago the most advanced tools were the chipped quartz handaxes of the type known as Acheulian; today there is the Cray-1 supercomputer and much else. It has all happened very fast.

However, the evolution of the genus *Homo* took place against the backdrop of the great climatic upheavals known as the Pliocene Drought and the Ice Ages. Everyone has heard of the latter; the former was a ten-million-year time ending some two million years ago in which the African rainfall was one-third of present-day normal. It is too much to ask of coincidence that the ascent of man just happened to be then. It is much easier to believe that these climatic crises gave a survival value to intelligence, and that in a mild, equable climate its rise might have been delayed.

The density of stars in our cosmic neighborhood is about one per six hundred cubic light-years. This means that with one ''good star'' to every 227,000, the nearest star that has harbored intelligent species is about five hundred light-years away. The nearest abode of an intelligent civilization would be at least that far, and possibly a good deal farther.

Most such planets could not be expected still to be active locales for intelligence. They would offer fossil beds, artifacts strange as from Easter Island, dusty-dry records in unknown scripts that crumble at a touch. To speculate on how many intelligent species or civilizations exist today, we need to guess how long they last. If the experience of Earth's primates is typical, then the average lifetime of families of intelligent creatures (who do not necessarily have advanced technical societies) is some three million years. This means only about four hundred such species, assuming one to a planet, in the Galaxy today. The nearest one would be about five thousand light-years off.

As for the lifetimes of civilizations, there we cannot even guess. They might be as short as a few hundred years or as long as a billion. Even with lifetimes of hundreds of millions of years, there may be no more than a few dozen existing today. Like wildflowers blooming after a desert rain, they could flourish in their moments of glory while never knowing of each others' existence. But if civilizations overcome the stresses of immaturity and youth and go on to reach advanced ages, then the Galaxy may be planted with them as thick as a forest. If billion-year cultures are at all widespread, the nearest may be within a hundred light-years.

Another instability as well is influential here. Any sufficiently advanced and long-lived culture may surely learn the arts of space colonization and interstellar flight. To such a culture, the Galaxy would appear as did the island-studded Pacific to the first seafaring Polynesians. It would be a fertile milieu in which small bands of settlers might set out for a nearby star, colonize it, and let their numbers grow. In later centuries the descendants would produce new starfarers who would repeat the process.

Cultures may perish, species become extinct; but the presumed hundreds of thousands of civilizations must surely have had the opportunity to test all possible ways of life, all philosophies. Surely one of them, if not many more, may have succeeded in using its intelligence to devise those forms and practices that would permit long cultural life together with interstellar colonization. Such a culture could be to the Galaxy what the Western Europeans have been to Earth.

It has been ten thousand years since men laid the groundwork for civilization by learning to domesticate the native plants and animals. In the millennia since, myriad cultures and ways of life have arisen. Yet the Western Europeans came up with the conbination of technical and ethical principles that swept the world. Advanced technology, freedom and justice (including the Communist versions of these ideals), the urge to explore and expand—this was the winning combination. Few are the nations that have rejected or been untouched by at least a part of this combination; its influence and its people have penetrated the remotest corners of Earth. Is it too much to expect that somewhere in the Galaxy there has arisen a similarly influential culture, and from its ancestral stock have sprung far-flung nations that have left their touch on the planets of nearby stars and perhaps within our very Solar System itself?

We thus must come to grips with the question of the existence of starships and of stellar visits. The concept of the starship is an old one; like the Martians in Nigel Kneale's "Quatermass," it is a name "almost worn out before anything arose to claim it." Yet it is quite a specific concept: a spacecraft capable of spanning the gulfs between the stars while supporting a community of onboard residents for the duration of the voyage. There has been much speculation about travel at close to the speed of light, as well as of keeping the travelers in suspended animation or freezing them for the duration. These speculations today are far from convincing. Nevertheless, there is a type of starship that even today can readily be envisioned.

This is simply a mobile space colony. The trend of space colonization may well lead to cosmic arks suitable for interstellar voyaging. Such an ark would be a colony in which self-sufficiency has been highly developed, allowing its people to live indefinitely by controlling the population and recycling materials.* For propulsion, thermonuclear power will be more than adequate as long as the speed is no more than a few percent that of light. If it speeds up to 5 percent of light speed and then decelerates at the destination, the total mission would require the energy obtained by converting 0.005 of the starship mass to energy, as in $e = mc^2$. In fact, if the starship propellant tanks contain a mass of hydrogen equal to that of the ark, and if this hydrogen is fused into helium via the reaction that powers the Sun, the energy released would be $0.007\,mc^2$.

This reaction actually proceeds so slowly as not to be practical or useful, but other thermonuclear reactions are speedier. These involve deuterium and tritium, the heavy isotopes of hydrogen, as well as the isotope helium 3. There is plenty of deuterium in the oceans—at least ten trillion tons of it—and it is readily extracted. A few million tons for a starship will never be missed. Helium 3 is rare on Earth, but that is merely because helium is rare. It is common in Jupiter and Saturn, and any culture that can build starships can certainly tap these sources.

Let us now imagine that some interstellar culture has done no more than reach this primitive level of technology that we envision today with our inadequate grasp of physics, but that we may well attain less than ten thousand years after inventing the wheel. It is well known in nature that species growth and extension continue as long as there are the resources to support further expansion. Indeed, one need only think of the world's small and remote volcanic islands—Hawaii, the Galapagos, Tristan da Cunha. All their native plants and animals (excepting those brought by man) are descended from ancestral colonists that survived the perilous voyage from the mainland: monkeys and other animals

---

* Which need not include the bodies of the dead. One indication of civilization is the respect shown the deceased; these could be cremated and the ashes stored in urns in a mausoleum. Even if ten thousand people journey through space for a thousand years, the amount of matter thus tied up would be only a few hundred tons, compared with the hundreds of thousands of tons that would be the mass of the starship itself.

floating on natural rafts of tangled tree-trunks, insects borne on the hurricane, seeds brought by birds of passage. With far safer and more convenient modes of transport, and with the energy and resources to support very great expansions, it is easy to imagine a space culture doing the same.

Animal species do not increase their numbers beyond the carrying capacity of their environment, and we must certainly expect that a long-lived intelligent culture will practice population control. (They would hardly be intelligent if they didn't.) In any planetary system that it colonizes, including the home system, its population will not exceed some optimum level. Even so, Eric Jones, of Los Alamos Scientific Laboratory, has reached a remarkable conclusion: If a stellar civilization practices population control but devotes even a small effort to building starships, it will produce a wave of colonization that will sweep across the Galaxy in no more than one-hundredth the Galaxy's age.

Jones' calculations involve four quantities:

The Optimum Population (O.P.) This is the level that can best be supported over long periods of time with the available resources and energy of a home planet, or of a planetary system, in the likely event that most of the population live in space colonies.

The Starship-Building Population (S.B.P.) This is the level, less than the O.P., at which the culture possesses sufficient resources to begin building starships. Smaller populations are assumed to be devoted to developing their planets and welcome immigration. Populations larger than the S.B.P. send out emigrants, but are assumed not to welcome new immigrants from pre-existing colonies.

The Growth Rate (G.R.) When the population is small, this is its yearly rate of growth. As the population rises, its growth rate slows, and as it approaches the O.P. it reaches zero population growth, in Jones's assumptions. The population never gets larger than the O.P. However, the G.R. is specifically the growth rate when the population is small.

The Emigration Rate (E.R.) This is some small fraction of the G.R. It represents the yearly rate at which inhabitants emigrate in starships.

The emigration does not proceed uniformly toward stars in all directions. Star systems suitable for colonization are distributed randomly, and at first the colonists move along preferred directions toward the nearest good stars. Some stars relatively close to the home world are not visited till a branch of the colonization wave turns in their direction. Then an expedition might travel inward to reach a star closer to the home world than was the original star it settled.

For example, let the S.B.P. be one-half the O.P.; that is, starship-building is a low priority and is not undertaken till a population is well established, well along on its growth to what will be its ultimate extent. Let the E.R. be only one-millionth the G.R., which we take as 0.005 or .5 percent per year. This means that interstellar flight is in no way a means to reduce population pressures or to delay the approach to a stable population. At the risk of anthropomorphizing the starfarers' motives, one could say that starflight is a marginal activity engaged in by a small number that are its devotees and enthusiasts. Even so, there is then an expanding wave of settlement that sweeps across the Galaxy at 1.5 percent the speed of the starships. Since the Galaxy is some 100,000 light-years across, if a starship travels at 5 percent of light speed this wave of settlement will cross the Galaxy in some ninety million years. This figure is less than 1 percent the Galaxy's age, 2 percent the age of Earth.

Much faster results follow if the culture regards starfaring with at least the importance eighteenth-century Britain gave to the colonization of America. That colonization was largely the work of voluntary emigrants such as might be attracted to space. This contrasts with the much larger later waves of settlement in America, which were driven by the Irish potato famine and by the oppressions of czars and emperors as well as by the demands of the Industrial Revolution. Even so, between 1700 and 1790 some 350,000 Englishmen came to North America. With England's population at twelve million in 1750, the E.R. is 0.0003 per year.

For the G.R., again take 0.005 or .5 percent per year. Let us further assume that the S.B.P. is only 1 percent of the O.P.; that is, a population waits only long enough to become well established in a new planetary system before it once again revives the founders' trade of starship-building. Since this culture takes starships seriously, let us also suppose that theirs are quite powerful and travel not at 5 but at a very fast 10 percent of light speed. Then the expansion wave goes at 16 percent of ship velocity. Within an evolutionary eye-blink of six million years, it has reached the farthest corners of the Galaxy.

We now must imagine that at some time in the past four billion years, some super-civilization (one of perhaps hundreds of thousands that have existed) launched just such an expanding wave of settlement. At some time along the way, like Fletcher Christian's *Bounty* mutineers discovering Pitcairn's Island, the outriders of this expansion discovered our solar system. There, at a convenient ninety-three million miles from its star, was a lovely green planet, which like Pitcairn at a much later date, proved to be totally untenanted and truly a fine place to settle and colonize. What can we say about this?

To begin, while a colonizing visit to Earth must be a rare event, the geologic strata could preserve the evidence, just as it preserves the remains of meteor craters. Impacts by large meteroids are indeed rare, but the ancient granite of the Canadian Shield has faithfully recorded them. Any colonizing visit would quite likely leave abundant remains of roads, pavements, structures, machines and equipment, even the bodies of the deceased.

Next, it is appropriate to note that just such evidence exists for major events in human prehistory. The archaeological remains of ancient towns in the Middle East, and even of Viking explorations in Newfoundland, are examples. An even more telling case is a 300,000-year-old campsite on the French Riviera used by a small band of the near-men known as *Homo erectus*. The anthropologist Henry de Lumley has found evidence for their use of poles in constructing shelters or lean-tos. He has found a firepit and evidence for the existence of specialized craftsmen at that early date. (A small area, well away from the fire, contains chips from stone tools such as would have been made by an artisan.) He has even found the fossilized remains of feces.

Finally, we recall that geologists, archaeologists, anthropologists, and other diggers have been very busy over the past century and a half. We now have a very complete and consistent understanding of the fossil record for at least the last half-billion years, and a partial understanding of the geologic record for two or three billion years further back. In addition to the excavations of scientists in the field, there are canyons, mountains formed by upthrusting great rocky blocks along fault lines, and the diggings of those who have built roads and cities.

In short, the geologic strata would have been patiently collecting any evidence of settlers from the stars for billions of years. We are quite familiar with that strata and have traced in them the evidence of even very localized and temporary events. And in all that vast archival storehouse, that natural library of Earth's history, there is not a single artifact that is indubitably from a source not of Earth. Three billion years and more, and not a single stretch of concrete, or steel beam, or fragment of an abandoned set of scientific instruments. No evidence for settlement or even for visits for the purpose of exploration. No counterparts of our own Viking, Pioneer, Venera, Surveyor, Lunokhod, and Apollo spacecraft of exploration, now resting on the surfaces of Mars, Venus, the Moon. As Fermi asked, "Where are they?"

Eric Jones in his article, "Interstellar Colonization," has made this point somewhat more poetically:

> If only one species arisen in the remote past colonised for a long time, the implications are profound. Consider the Lornax, a gentle race of camel-like but omnivorous creatures from the distant world of Lorna. A few hundred million years ago the first Lornax colonists arrived in the Solar System weary from the long voyage. Quickly finding earth and an ideal seaside landing site on the west coast of southern Pangea, the Lornax settled in for a long stay. The fall of night found the first settlers enjoying the primitive warmth of a driftwood fire; remarking how the windswept dunes evoked images of the near-mythical homeworld. In the morning, they vowed, they would catch more of the exquisitely flavoured lungfish.
>
> This did not happen. We are the decendants of lungfish.

In all fairness, one should not ignore the people who believe, with Erich von Daniken, that the cultures of prehistory, the civilizations of antiquity, could not have arisen without intervention, divine or extraterrestrial. However, most of the great works that are presented as proof fall into the categories of positional astronomy or monumental architecture. Regarding ancient astronomy, any high priest with a bent in that direction could have done quite a creditable job with even a crude transit instrument,* particularly if his work was part of a centuries-old tradition. Regarding ancient architecture, there are people who forget (or don't want to remember) that in those times there really were mighty kings with the power of life and death, who could cause great cities and monuments to rise in their names. In none of this is there convincing reason to think that humanity has ever had help from beyond the stars.

Yet the matter does not end here. It is quite possible for nonterrestrial visitors to come to Earth without leaving evidence in the geologic record. This possibility lies behind the idea that some of the UFOs are visitors from space.

There is a vast literature on unidentified flying objects, but very little of it is persuasive. It is far from rare for even experienced observers to see strange lights or flying objects, which turn out to be weather balloons or airliners, or planets such as Venus seen under unusual conditions. If all UFO sightings could be explained this way (and indeed most can), then there would be no more to the matter of UFOs than that people sometimes respond in strange ways to unfamiliar things. The problem is that when one filters through the reports, there is a hard core of good observations that cannot be explained.

A review of some of the most unusual UFO encounters will illustrate the problem. On July 17, 1957, an Air Force RB-47 reconnaissance aircraft was followed by a UFO for some ninety minutes and for over seven hundred miles as it flew across the southcentral U.S. The aircraft was equipped with electronic countermeasures gear and was manned by six officers, three of whom were radar specialists. At various times the cockpit crew saw the UFO as an intensely luminous light, and it was also detected by the plane's electronics equipment. It was also followed by ground radar. Several times the UFO disappeared and then reappeared, and these events were seen simultaneously by eye as well as by ground and air radar. At one time the UFO was seen at an altitude of twenty thousand feet

---

* A transit instrument is a tube or pole along which one sights so as to view a star or planet. It is equipped with marked circles to measure the pointing directions. A sixth-grader today can make one and use it, and the ancient Near East priesthoods were a good deal smarter than that.

when the RB-47 was at thirty-five thousand. The captain received permission to dive at it. As he approached twenty thousand feet the object blinked out, disappeared from the ground radar scope, and disappeared from an onboard radar monitor, all at the same time.

On August 13, 1956, a radar at Bentwaters, England, tracked something traveling at four thousand miles per hour westward, at four thousand feet altitude. At the same time, control-tower operators reported a bright light passing overhead toward the west, and the pilot of a C-47 aircraft, four thousand feet over the airfield, saw the bright light streak westward underneath him. Someone at Bentwaters station then phoned the second radar station at Lakenheath: "Do you have any targets on your scopes traveling at four thousand mph?" The Lakenheath station then detected a stationary target at twenty thousand feet altitude, which suddenly and with no buildup in speed went north at six hundred miles per hour. It made several sudden stops and turns, which were seen by two radar sets and three ground observers. After about forty minutes of this, an RAF Venom night fighter was vectored in for a close look. The pilot detected the UFO with his radar and locked his guns on it. Suddenly, with a quick circling movement, the UFO moved behind the Venom; the pilot took evasive action but could not shake it. All this was seen on ground-based radar as well. When the fighter flew off toward its base, the UFO followed only a short way. Then it once again became stationary before resuming its flight to the north.

On January 20, 1967, near Methuen, Massachusetts, three people were driving through a lonely area. Reaching the top of a hill they suddenly came upon a straight string of glowing bright red lights, apparently a few hundred feet away. The driver slowed the car and proceeded toward the lights. When almost broadside to them, the object to which they were attached swung around to reveal a new appearance. Four distinct lights formed a trapezoid, two red ones for the top and two white ones for the base. The driver pulled over to the side of the road directly broadside to the UFO, which now seemed to be lower and nearer; the engine was idling with lights and radio on. Then suddenly the engine, lights, and radio failed completely. The driver tried to start up the car but the engine gave only a low "moan." He shut off radio and lights and tried again, but still could not. The driver's side window was open, but there was no sound. Then the object began moving and then shot away at high speed. The driver then was able to start the car and the lights and radio worked perfectly.

In addition, UFO sightings extend back into history. An example has been reported by Carl Sagan in "Communication with Extraterrestrial Intelligence":

> This is the report of a monastery clerk in Northern Russia addressed to a high dignitary of the Russian Church who reports that on August 15, 1663, there was a visitation of the earth between ten and twelve hours from the clear skies. A sphere appeared, about forty meters in diameter; from the lower part two rays extended earthward and smoke poured from the sides of the vehicle. The body disappeared and reappeared again, again disappeared and reappeared, changing in brightness in the course of these peregrinations. The phenomenon appeared over a lake and lasted for an hour and a half. At the place where the sphere touched the water, a brown film appeared, resembling rust. The phenomenon was witnessed by two groups of people. Some watched it from the church; others from a boat which happened to be in the middle of the lake.

Such reports as these certainly sound like extraterrestrial visits. Yet such an explanation merely fits the spirit of the times; no more. We can conjecture that some UFOs are from outer space; we cannot readily prove they are not. Yet in earlier centuries the same would have been true for other explanations of UFOs, fitting the spirit of those times: angels' haloes, temptation of the devil, the spirits of departed kings, celestial chariots heralding the return of the Risen Christ, or the souls of the dead. Such explanations cannot be disproved any more readily than can the extraterrestrial conjecture.

What's more, this conjecture runs into inconsistencies. The idea that UFOs are observing us arose after World War II because of the peculiar resemblance of some of their activities to wartime reconnaissance aircraft. The similarity is really quite striking if we imagine, say, a flight of British "Mosquito" aircraft sent to reconnoiter a German target such as the battleship *Tirpitz*. The planes would fly in formation, slipping in and out of clouds, then venture to approach the *Tirpitz* while keeping a respectful distance. One pilot might swoop in quickly for a close encounter, to get the best photography. If he met too much anti-aircraft fire he would swiftly fly away; if the German defenses proved too strong, the whole formation would fly off at top speed. There are a number of UFO reports of this type.

Yet only twenty years after the sinking of the *Tirpitz,* the Air Force had the SR-71. Flying at eighty thousand feet and at two thousand miles per hour, it could scan the entire U.S. in only three passes with side-looking radar. Such performance is typical of reconnaissance today, yet is very untypical in the world of UFOs.

There is another and very telling problem with the UFOs; namely, they have not been detected by space surveillance. Since the 1950s the U.S. and its allies have set up an impressive array of radar networks to track objects in space. These include the Ballistic Missile Early Warning System, with powerful over-the-North-Pole radars, as well as the Air Force's Space Detection and Tracking System. This latter system routinely tracks objects as small as an astronaut's glove, which floated out of a Gemini spacecraft. In addition there are spaceborne reconnaissance satellites. One of them tracked a great meteor on August 10, 1972, which grazed Earth's atmosphere and barely missed impacting the surface. Had it hit, it would have struck with the force of the Hiroshima atom bomb. It was a most rare astronomical object; yet it was tracked from space. The same cannot be said of UFOs.

For a time in 1960 it appeared that our radar networks had indeed detected a visitor from space. On January 31 a Navy network known as Dark Fence detected two passes of an unknown space object. It was nineteen feet long by five wide, in a nearly polar orbit ranging from 134 to 1,074 miles. The commander of Dark Fence personally reported this to Admiral Arleigh Burke, Chief of Naval Operations, who took the news to President Eisenhower. The Soviets declared it was not one of theirs; U.S. specialists declared it was not one of theirs. It turned out that the intruder was an Air Force satellite, Discoverer V, launched the previous August. It had carried a retro-rocket to return a capsule to Earth, but the rocket fired in the wrong direction. The result was a higher, unpredicted orbit.

The fact that no similarly unknown objects have been tracked raises a real problem for UFOlogists, especially since UFOs have frequently been detected on radar. One could argue that alien spacecraft can make themselves radar-invisible. But why should they do that when they are out of the atmosphere, yet make themselves radar-visible when inside it?

Even very good UFO reports can be explained without resorting to the "we-are-being-visited" idea. Many apparent radar UFO sightings have actually been due to effects known as anomalous propagation, in which radar signals are channeled in the atmosphere to produce what are in effect radar mirages. In the words of one specialist in this field, "There is now abundant evidence that the atmosphere will effect radar propagation in almost unbelievable ways and produce virtual [apparent] targets which have apparently fantastic maneuverability." Such images gave rise to a UFO flap in midsummer 1952 at Washington, D.C. In response to these strange radar echoes, interceptors from

Andrews Air Force Base scrambled to seek the sources. Yet they were no more than false radar returns.

One well-explained UFO report could virtually have been taken from the movie *Close Encounters of the Third Kind*. A child, going to the bathroom in the middle of the night, turned on a light and woke his parents. Suddenly the light went out. The father got out of bed to investigate and happened to glance out a window. What he saw was a pulsating, reddish glow that moved irregularly over the sky, then faded out.

The explanation? A blown transformer had knocked out power just as the child turned on the light. The father, his eyes dark-adapted from sleep, caught the bright light full in his eyes. The result was an after-image, which drew his attention as he passed the window.

Some of the *Close Encounter* cases are more subtle. Imagine someone driving at night along a lonely road, his attention wandering. Suddenly he tops a hill and sees the rising full moon, copper-colored, hazily shining through distant mist. If he does not recognize what he sees, he may be startled, and if of a suggestible temperament he may even panic. Then he will likely say that the light is brighter, larger, nearer than it really is. This sort of thing is actually quite common in UFO lore, as when witnesses fail to recognize meteors blazing a hundred miles away and report a UFO that passed ''just over the tree-tops.''

What of the failure of auto ignitions, radios, and the like? I had a personal experience of that type once. I was driving along a nearly deserted stretch of Interstate 10 north of Tucson, Arizona. It was a bright, hot September day and my mind was only partly on the road. Soon my attention was diverted by a passing train. As I watched it, my car meandered over to the shoulder. Startled, I hit the brake and turned not knowing the car would go out of control. By the time I regained control the car had spun completely around and was coasting backwards; the ignition was dead. When I recovered my wits, I started up the engine and drove off, but in a far from calm state of mind.

Suppose it had been not a train that distracted me, but a coppery and unfamiliar-looking Moon at night (which vanished behind clouds a minute later). Were I of a suggestible turn of mind, the result might not have been just a case of my being a damn fool at the wheel. Instead, it could have been a classic Close Encounter of the Second Kind.

Still, such explanations leave a residuum of unexplained cases, of which Lakenheath and the RB-47 case are only two. What are we to say when multiple witnesses (possibly with instruments) calmly and with great care make mutually corroborating observations of a strange phenomenon in the sky? Here indeed we may have something new to be to be learned. It may take decades or even centuries, but the surprising explanation will indeed be found.

A very good case of just this type took place near sunset on June 18, 1178, near Canterbury, England. Here was not a UFO, but a strange event involving the Moon. As recorded in the medieval chronicles of Gervase of Canterbury,

> A marvelous phenomenon was witnessed by some five or more men who were sitting there facing the moon. Now there was a bright new moon, and as usual in that phase its horns were tilted toward the east; and suddenly the upper horn split in two. From the midpoint of this division a flaming torch sprang up, spewing out, over a considerable distance, fire, hot coals, and sparks. Meanwhile the body of the moon which was below writhed, as it were, in anxiety, and to put it in the words of those who reported it to me and saw it with their own eyes, the moon throbbed like a wounded snake. Afterwards it resumed its proper state. This phenomenon was repeated a dozen times or more, the flame assuming

various twisting shapes at random and then returning to normal. Then after these transformations the moon from horn to horn, that is along its whole length, took on a blackish appearance. The present writer was given this report by men who saw it with their own eyes, and are prepared to stake their honor on an oath that they have made no addition or falsification in the above narrative.

What was it? To the lunar geologist Jack Hartung, the clue was the "flaming torch" which spewed out "fire, hot coals, and sparks." So might a meteoroid impact appear, flinging out white-hot rock to a great height above the Moon. The statement that "the upper horn split in two" means the impact could have occurred only in the region of lunar latitude 45° north, longitude 90° east. In fact, at 36° N by 103° E is a large, very bright, very fresh-looking crater, which from geological evidence appears to be among the most recently formed large craters on the Moon. It is named Giordano Bruno, after a sixteenth-century scientist who, ironically, argued in favor of the plurality of inhabited worlds. Had he lived a few centuries earlier, he would have had the unique honor of having his crater form after his death.

If even the best UFO reports are to fall before similar unusual explanations, then we can rule out the idea of interstellar visits in the present as well as of interstellar colonizations here in the remote past. Still, there is the possibility that advanced cultures engage in signaling or communication. In the past two decades the interest in SETI (Search for Interstellar Communications) has blossomed considerably. There may soon be a formal NASA program in this area, to be run by Dr. John Billingham of Ames Research Center. Thus far, Congress has voted no funds for the work, but the bureaucratic machinery is already in place. Billingham and his associates already have office space in Building 204 at Ames, and outside that building is clear evidence that SETI may soon be just another government project. There is a parking space marked CHIEF, SETI PROGRAMS.

The SETI experiments conducted or proposed to date have used the techniques of radio astronomy, which some people have suggested will turn out to be a primitive, quickly abandoned technology at the interstellar level. Advanced cultures could use powerful lasers, neutrino beams, gravity waves, or whatnot. However, any communications channel must compete for detection against the overwhelming power of the nearby star. It is by this comparison that radio transmission quite literally shines; it can easily be made far more powerful and readily detectable than the competing natural radio emissions. In any case, even if new communications modes come into use, the old ones may still be valued. Despite the efficiency of the telephone to transmit messages, we still also use a system that dates to the ancient Sumerians—the post office. Nor have we abandoned the institution that a century ago was as commonplace as sarsaparilla and mansard roofs, the telegraph. Quite the contrary; Western Union has moved with the times and has lately been building and launching its Westar communications satellites.

The fact that an extrasolar civilization might beam radio signals to us does not necessarily mean that the signals would carry information that we could decode. The most efficient way of encoding information is known as Shannon coding, after the theorist Claude Shannon of MIT. To one who lacks the means of decoding, a Shannon-coded signal is indistinguishable from random noise. An example of this is a hologram, which encodes vast quantities of information about an object's three-dimensional image. Such images can be reconstructed using laser light. Yet if one seeks the images by examining the hologram under the microscope, all that will be seen is a nearly random pattern of speckles resembling a bad case of snow on a TV screen. Still, there is a clear means of identifying artificial signals from space. Unlike natural signals, they will use only a very narrow range of radio

The crater Giordano Bruno, whose formation may have been witnessed by five men near Canterbury, England on June 18, 1178. (Courtesy NASA and Lunar and Planetary Institute)

frequencies. This is because for a given signal power, the transmission can be detected much more strongly and at greater distances if it is concentrated and not spread out in frequency.

Even the mere detection of an artificial signal, however devoid of content, would be the astronomical equivalent of the California gold rush. The first detected signals would play the role of nuggets of gold, which point to the existance of a rich mother lode. These early signals would be regarded as ''acquisition signals'' to direct our attention to the particular part of the sky. We then would dig down with receivers of increasing sensitivity, till we hit the lode: signals generated by that culture for its own purposes. These might be similar to our own TV transmissions and missile-detecting radars.

This does not mean tuning in the celestial equivalent of an ''I Love Lucy'' rerun; it takes ten thousand times better radio sensitivity to produce a good TV picture than merely to detect its carrier signal. Still, even without detecting any information content in such interstellar signals, careful tracking and study would tell us many things about the distant planet and its culture. Merely by studying such signals with standard techniques of radio astronomy, we could learn:

The complete orbit of the home planet, as well as its size and rotation rate.

The existence and nature of large natural satellites such as the Moon.

The structure of the atmosphere, including the presence of an ionosphere and of winds which bend transmitting antennas.

The existence of daily broadcast schedules keyed to the planet's rotation.

The size of the transmitting antennas, which would give clues to the sizes of structures on that planet.

A complete map of the locations of the transmitting stations.

Various cultural inferences concerning the civilization. It might be noted that well-defined regions of the planet (nations?) follow different conventions for frequency allocations, broadcast

219

schedules, and signal strengths. Comparison of these conventions could even suggest that some regions are wealthier or more populous than others.

With such rich rewards in store, it is little wonder that a number of radio astronomers have devoted much effort to seeking the powerful acquisition signals. The first such search, Project Ozma in 1960, generated much publicity (including a cover story in *Newsweek*) but few results. More recent searches have proceeded with much less fanfare but have examined many more regions of the sky. As of 1978, the list of such surveys is already somewhat lengthy.

The accompanying chart shows that SETI is rapidly becoming an important activity in the community of radio astronomers. The trends have been to examine more and more stars, with better and better equipment. To date some fifteen hundred stars have been studied. The first searches used simple equipment, but the most recent ones have used such great radio telescopes as the one-thousand-foot Arecibo antenna and the three-hundred-foot transit telescope at the National Radio Astronomy Observatory. The most recent work has indeed been keyed to detecting signals that span a very narrow range of frequencies; the telescope then acts like a TV receiver with 65,536 channels. The chart's last column, "Sensitivity," is deceptive in its simplicity. The best sensitivity to date, 4×10^{-27} watts per square meter, would suffice to detect a forty-watt bulb at twenty times the distance to Pluto.

Some of these surveys examine entire galaxies. This search strategy is akin to looking for life on Earth by observing the whole of North America at night with a telescope, and looking for the lights of cities. If any star in a galaxy boasts a civilization that has established a sufficiently powerful beacon, we could detect it and say there was intelligent life there. Such observations could well yield the first detection of artificial signals. The problem of SETI is to point a radio telescope in the right direction, at the right time, and to detect the right frequency with sufficient sensitivity. We already know how to scan simultaneously a broad range of frequencies so as not to miss the right one, and it should become possible soon to devote full time on a large instrument for this work. Pointing it at a nearby galaxy could also solve the aiming problem, since the signals would not be coming from just anywhere but rather from the limited region of the galaxy only.

Such galactic transmitters might be called Ozymandias method of signaling, after the poem by Shelley:

> My name is Ozymandias, King of Kings;
> Look on my works, ye mighty, and despair!

We can imagine that a culture wants to make its presence known across the Universe, as well as to achieve a kind of cosmic immortality by continuing to signal automatically long after the culture itself has gone to dust. Such motives are not unknown on Earth, and Ozymandias could be a type of pyramid-building on a scale never dreamed of by Cheops.

The interesting feature of Ozymandias is that it need not be the work of some mythical super-civilization that has tapped all the energy of its sun. Suppose that Ozymandias is to illuminate a distant galaxy with the power of 4×10^{-27} watts per square meter, the sensitivity limit of the best experiments to date. To be specific, let the target be a circle 100,000 light-years in diameter, the approximate size of the Milky Way. The nice feature of Ozymandias is that it then doesn't matter if

SUMMARY OF SETI EXPERIMENTS THROUGH 1978

Most stars examined have been within 100 light-years. An arrow indicates the search is continuing. No observation to date has yielded evidence for artificial signals.

Date	Investigators	Diameter (feet)	Frequency (hertz)	Total bandwith examined (megahertz)	Resolution (hertz)	Target	Sensitivity (watts/meter^2)
1960	F. D. Drake	85	1.42	0.4	100	2 stars	10^{-21}
1968	V.S. Troitskii et. al.	45	0.927	2.2	13	12 stars	2×10^{-21}
1970→	V.S. Troitskii	Dipole	1.875			Omnidirectional	
		Dipole	1.0				
		Dipole	0.6				
1972	G. Verschuur	140	1.42	20.0	7,000	10 stars	1.7×10^{-23}
		300	1.42	0.6	490	3 stars	5×10^{-24}
1972→	B. Zuckerman, P. Palmer	300	1.42		3,000	602 stars	5×10^{-24}
1972→	S. Bowyer, M. Lampton, J. Welch, D. Langley, J. Tartar, A. Despain	85	Variable	20.0	2,500	Semirandom	10^{-21}
1973→	N.S. Kardashev	Dipole				Omnidirectional	
1973→	F. Dixon, D.M. Cole	175	1.42	0.38	20,000	Area search, $7° < \delta < 48°$*	2×10^{-21}
1974→	A.H. Bridle, P. Feldman	150	22.2		30,000	500 stars	8×10^{-25}
1975→	F.D. Drake, C. Sagan	1,000	1.42, 1.653		1,000	Several galaxies	4×10^{-25}
		1,000	2.38		1,000	Several galaxies	4×10^{-27}
1978	P. Horowitz	1,000	1.42	0.001	0.015	185 stars	
1978	J. Tartar, J.N. Cuzzi, D.C. Black, T. Clark	300	1.653	0.33	5	201 stars	10^{-23}
1978	M.A. Stull, F.D. Drake	1,000	1.653	4.2	0.5	10 stars	10^{-26}

*δ, Declination

the target galaxy is near or far. To signal to a galaxy 10 billion light-years off, the system merely focuses the beam to reduce its spreading.

The transmitting power of Ozymandias then is 2.8 10^{-15} watts. That's 100,000 times the level of a power satellite, but only ten-trillionths the power output of the Sun. If Ozymandias were powered by an enormous powersat, it would be a square twenty-seven hundred miles on a side. This is quite a large structure, but if we can seriously propose powersats the size of Manhattan Island, our descendants may well be able to envision one the size of the Moon.

If we were to build an Ozymandias transmitter, we could fulfill the suggestion of author Lewis Thomas when asked what message he would send for interstellar signaling: "I would vote for Bach, all of Bach, streamed out into space, over and over again. We would be bragging, of course, but we can tell the harder truths later."

An immense array of transmitters like these might be used in the Ozymandias program for signaling distant galaxies. (Courtesy Rockwell International)

Ozymandias could indeed be a creation of Earth itself, in a century or so. Its cost would be close to a quadrillion dollars, which is a hundred times more than today's Gross World Product. Still, economic growth has its little ways of overcoming such matters. It is worth remembering that in the last century, one million dollars played roughly the same role as does one billion dollars today. The first billion-dollar federal budget came in 1889, and it will be surprising if the 1989 budget is much below a trillion dollars.

We have not yet detected anyone else's Ozymandias, though as yet we have not searched very hard. Still, the lack of success in all searches to date is sufficient to give one pause. In the past thirty years there have been any number of astrophysical phenomena predicted to exist, searched for, and found. There have even been important discoveries like quasars and pulsars, which were entirely accidental. (Pulsars were at first thought to be a cosmic time-standard broadcast service, but this theory was soon disproved.) In all this time there has been a predicted effect that has not occurred: transmissions from other civilizations.

This, then, is the Fermi paradox: We expect to find some clear evidence of the existence and pervasiveness of intelligence as an occurrence in astronomy, and we do not. Not in the rocks, nor in our skies, nor yet with radio telescopes.

We have not succeeded in resolving this paradox; therefore, we must look further. While doing so, one possible answer to this paradox should be kept in mind: We are alone; we are unique, there is none like unto us in this good green garden of our galaxy.

CHAPTER 12

An Interstellar Hang-up?

Where are the extraterrestrials, indeed? The answer will not be found by further belaboring of the sciences; not in astronomy, nor in planetary studies, nor yet in biology. In all honesty, one must acknowledge that at this point there no longer is firm understanding to guide us; we have come down a path that ends in mystery.

Yet it could hardly be otherwise. The search for our galactic neighbors may tantalize future generations; it may serve as the wellspring of much science and art in the space communities of our descendants. If so, then it could hardly admit to a simple and clear resolution at so early a date as the late 1970s.

Still, if we cannot know, at least we may speculate, and speculate we will. What we are now led to pursue is the anthropology of extrasolar cultures. This is a science whose very subject matter is not known to exist. Its pursuit is reminiscent of that of Thales and Democritus, who speculated on the nature of atoms and matter in days when the Parthenon was newly minted. Still, there is no alternative. There are only so many reasonable means of resolving the Fermi paradox:

Extraterrestrial civilizations do not exist. Yet the last chapter has given good reason why they may have arisen in up to 880,000 locales, then spread across the Galaxy. Or,

Such civilizations perish by their own hand before they succeed in gaining a galactic foothold. This explanation deserves attention because it would mean there is an interstellar hang-up: that intelligence does not promote long species life. Or,

Such civilizations exist, but for reasons of their own have not and will not make their presence known to us.

This chapter thus will consider the second and third explanations.

Is there perhaps an interstellar hang-up? This idea, that civilization may bring its own destruction, is not new but is merely a variant on a theme in philosophy over two centuries old. It was Jean Jacques Rousseau who introduced this theme in 1749 with his essay in the *Mercure de France* on the subject, ''Has the progress of science and the arts contributed to the corruption or purification of morals?'' In that essay, as well as in such influential books as *Émile* and *The Social Contract,* Rousseau argued that prior to the advent of civilization, people had been happy and virtuous, but that modern times had sown corruption and misery and had brought the exploitation of the many by the few. He then went on to argue that by means of appropriate reforms, mankind could be returned to what would be more nearly like his original state of natural goodness.

Rousseau's work was founded on a fallacy. In his day there was no science of early or primitive man, no way of knowing his beau ideal never existed. His Noble Savage never was found. Yet his influence has been comparable to that of Christ or Mohammed. With elaborations, his doctrines were taken over by Marx and then Lenin. In our own day they echo in such notions as the perfectibility of man and in the urging that we renounce technology and return to the land. They also echo in the exalting of emotion over reason and in the idea that civilizations sow the seeds of their own destruction.

Among the most significant insights that were denied to Rousseau was Darwin's theory of evolution by natural selection. In any animal population, including man, individuals are neither created equal nor endowed with identical traits. They differ, each from the other; there is variation among them. The variation involves physical traits such as size, speed, ability to withstand hunger, etc., but it also involves behavioral traits. Some animals are more aggressive in the chase, more sensitive in ability to detect prey, more assertive in the scramblings for food or females.

This variation ensures that some will best be fitted to survive and to pass on their genes, while others, less fortunate, will fall to illness or to the predator. However, conditions do not stay the same, and traits that may make species fit in one century may prove much less favorable if the situation in which they live should change.

When conditions change, the variation within a species may endow certain of its members or groups with traits that by chance are better suited to the new conditions. It will then be those that are favored and which will be the most fit. In this way, the existence of variation allows species to evolve.

The same is true of cultures. Even nuclear war need not prove a destroyer of intelligent species, for by the cultural adaptation of building shelters and preparing against the threat, many would survive. A nuclear war might well resemble a severe forest fire. In such a fire, there are always some trees with thick bark that survive, as well as brush and seeds that escape destruction. Soon after the fire, the burned-over area begins to show signs of life. As the decades progress, there is a forest succession, as one by one the old species return. Within a century or less, the succession has reached maturity, and the new forest is virtually indistinguishable from what had existed before the fire. The life of a forest can be damaged severely in fire, yet the forestland will not be turned to desert or blasted wasteland. Fragile in the short run, life is tough and resilient in the long, and acts to re-establish itself in familiar patterns in surprisingly short times. Some people have spoken of dangers from overpopulation or ecological upheaval, but biologically speaking, such threats appear unconvincing. The theory is predicated on the idea that an intelligent species is an undifferentiated and passive mass, totally at the mercy of its surroundings, with all members succumbing together.

In fact, any species resembling man would possess an incredibly rich diversity of individual and **225** cultural behavior patterns. In a time of challenge or stress, at least some of these patterns would enable them to develop, in advance, means to meet the challenge; as a biologist would say, they would be preadapted. Those cultures that truly wished to survive then could use their intelligence to learn from and adopt the ways of the preadapted ones. By a process of cultural selection, far more rapid than Darwin's natural selection, the best new cultural forms would be proved by experience and made available to all. This is not to deny that individuals and cultures would face wrenching readjustments and difficult, often stark choices. Rather, this viewpoint denies that such cultures would perish with the old rather than adapt to the new, for such stolid inflexibility would hardly mark a species as intelligent. On our own planet, a growing worldwide trend to birth control and limited family size illustrates this point and demonstrates anew mankind's cultural adaptability.

What of an eco-disaster or climatic upheaval? It is hard to imagine one more pervasive or far-reaching than the Ice Ages. Still, humanity survived that time quite nicely, thank you, even without modern technology, although the same was not true of the woolly mammoth or sabre-toothed tiger.

Short of a runaway greenhouse or runaway glaciation, as in Chapter 2, the worst disaster to Earth would be one that would bring widespread temperature increases to above 100°F. At such levels many animals, including man, would become infertile. Among the mammals and reptiles, it is usual for males to produce sperm in their testes, but sperm are quite delicate and sensitive to temperature. The testes must remain reasonably cool if the sperm are to avoid damage, for their proteins are subject to being denatured or broken apart by heat. In man injury would occur even if his testes were at normal body temperature, 98.6°F. It is to achieve the necessary coolness that man and other mammalian males carry their testes in a scrotum. The scrotum then dangles free and can be cooled by passing breezes.

War, overpopulation, environmental upheaval can all threaten the continued existence of species or civilizations, but none are inevitably the agents of doom, and in all there are opportunities for intelligence to avoid them entirely or to cope. Though not obvious today, eventually we may take these ideas for granted. Significantly, many people already do so and assume that intelligence even will overcome another threat: pestilence or disease. Yet human history shows this problem to have been by far the greatest threat of all, as when the Black Plague killed one-third of Europe's people in the 1340s.

Another possibility that can easily be challenged is that if a civilization exhausted its high-grade ores and mineral resources prior to lapsing into a Dark Age, it could never again recover. Yet the steel and other materials would not have vanished into thin air; rather, the nature of the "rich ore bodies" would have changed. Instead of digging for iron ore in some version of the Mesabi Range, one would dig for fabricated iron and steel at the sites of what formerly were great cities and harbors. A single wrecked ship of fifty thousand tons would satisfy for a century the need for cast steel of a civilization at the level of Germany in the mid-1830s, which then counted thirty million people.

Moreover, we must remember that even if one intelligent species should become extinct, there may be others waiting in the evolutionary wings, ready to step forward. The rise of intelligence could take place more than once, with the earlier, failed species sending their warnings to the future in the fossil record.

Is it then truly reasonable to say that if civilizations prevail for long times, they indeed will pursue interstellar flight, interstellar colonization? Perhaps after a phase of technical advance, such cultures lose interest in exploring or expanding.

Michael Hart, whose calculations were discussed in Chapter 2, calls this idea the Contemplation Hypothesis, the idea being that such cultures pursue spiritual contemplation, not science. In a classic paper titled ''An Explanation for the Absence of Extraterrestrials on Earth,'' he remarked,

> This might be a perfectly adequate explanation of why, in the year 600,000 B.C. the inhabitants of Vega III chose not to visit Earth. However, the Vegans of 599,000 B.C. could well be less interested in spiritual matters than their ancestors were, and more interested in space travel. A similar possibility would exist in 598,000 B.C. and so forth. Even if we assume that the Vegans always remain uninterested in space travel, there is still a problem. The Contemplation Hypothesis is not sufficient to explain [the absence of extraterrestrials] unless we assume that it will hold for *every* race of extraterrestrials, at *every* stage in their history after they achieve the ability to engage in space travel. That assumption is not plausible, however, so the Contemplation Hypothesis must be rejected as insufficient.
>
> The same objection, however, applies to any other proposed sociological explanation. No such hypothesis is sufficient to explain [the absence] unless we can show that it will apply to every race in the Galaxy, and at every time.

There is one more possibility that deserves note: the zoo hypothesis. It holds that Earth has been set aside as an interstellar wilderness preserve akin to Tanzania's Serengeti Plain or South Africa's Kruger National Park. Here our evolution would proceed, our planet's species interacting very little with the cosmic park-keepers.

This is a rather romantic notion, yet it has problems. We have our game preserves, true; but there is also such a thing as poaching on game preserves. To guard against the wily poacher requires considerable expense and effort in the administration of parks like the Kruger; in the Serengeti, less well guarded, poaching is a serious problem. In the cosmic context, ''poaching'' would mean the settlement or colonization of a world, in the face of efforts by a Galactic Empire to restrict or prevent such acts.

It might be that if another culture were to settle on Earth, it would do untold damage to the species here. So have our own colonists of oceanic islands done harm by bringing in rats or goats that have destroyed native birds or forests. Since planets like Earth appear to be rare, star-farers might have a strong reluctance to take the risk of damaging our life-forms.

Is this the answer, then? Is there a system of galactic ethics whereby Earth may have been discovered but thereafter left strictly alone, with not even the asteroids colonized in view of the ever-present temptation that then would exist to venture earthward? Yet this demands a resistance to temptation, a forebearance and devotion more that of angels than of men. The existence of such worlds could not be kept secret, but would prove a continuing enticement to would-be settlers or planners of expeditions. A galactic culture could seek to guard such planets from settlement but space is vast and difficult to patrol, and there would be the age-old problem: Who guards the guardians? Even *Volkspolizei* on the Berlin Wall, an elite and carefully screened group, have been known to escape to the West.

Beyond this, there is a simple biochemical rule that could give the starfarers the ability to have their cake and eat it too, to settle and colonize at least small parts of a world with no fear of destroying the local species by the celestial counterparts of rats or goats.

Many biochemicals exist in two structural forms, identical in all respects save one: symmetry. Such molecular structures are akin to a glove or shoe in that they characteristically can be described as **227** left-handed or right-handed. Other molecular forms are more nearly like an auto, which may have the steering wheel on the right (as in England) or left (as in America), the rest of the car being symmetrical. Even with such restricted asymmetry, though, we still describe the molecule as left-handed or right-handed.

The "handedness" of a molecular structure is established by determining its relationship to a standard molecule, glyceraldehyde, which exists in two forms. These are designated D (dextro, or right-handed) and L (levo, or left). All of life's asymmetric molecules can be traced back and compared to this standard, which produces a very interesting result. With only trifling exceptions, all sugars, including those found in nucleic acids, follow the D-standard. All amino acids, and hence all proteins, are L-standard.*

Why? Because this type of arrangement best suits the needs of enzymes, which govern virtually all of life's chemical processes. Enzymes are enormously complex chemical structures, which must fit carefully to bio-molecules, as a key fits a lock. If enzymes can deal with only one of two possible asymmetric structures, their tasks are greatly simplified. An enzyme attempting to deal with the other structure would be like trying to open a door using the key upside down.

This biochemical asymmetry furnishes strong evidence for the argument that all terrestrial life has evolved from a common ancestor. There is no convincing reason why life should have been based on D-sugars and L-amino acids; most biochemists regard this primitive selection as a matter of chance. The first successful living cell followed this rule, presumably, and so it has been passed down to all, from giraffes to geraniums. Nor was there a second successful origin of life, based perhaps on L-sugars and D-amino acids. Had there been, we might now find Earth populated by two ecologies, each consisting of life forms built according to these different rules. Significantly, life forms with one system would not be able to eat forms based on the other.

Suppose a D-symmetry cat were to eat an L-symmetry mouse. Once in the cat's stomach, the unfortunate rodent would be digested in the usual way, its body broken down to simpler substances. The sugars and amino acids would pass through the cat's intestinal walls into the bloodstream, and then be delivered to the cells. There the problems would begin. The cells' enzymes would not recognize the mouse's biochemistry. The cat would fail to receive needed nourishment and would very likely suffer a severe allergic reaction. Its enzymes might find themselves stuck to the mouse's molecules, as a wrong key will stick in a lock, and these enzymes could thus be poisoned. Some products of digestion of the mouse might build up in the cat's bloodstream, with no easy way for their removal. In a short time one would find that while the cat had killed the mouse, the mouse then killed the cat.

If there truly were such parallel but mutually inedible communities of life, over geologic time some species might develop special enzymes that would allow them to cope with the topsy-turvy biochemistry. The species' eventual prey would also learn to cope with the growing attentions of the predators. Instead of a sudden ecological shock, there would be plenty of time to evolve prey-predator relationships that would allow both to coexist, even in symbiosis. Naturalists know that in well-established ecologies that have evolved over a long time, the role of predators is not that of agents of wholesale slaughter. Instead, they take mostly the weak, the old, the unfit, and thus are agents of natural selection.

So if a starship commander were to seek assurance that his people could safely land, all he would

---

*Some D-amino acids are found in the cell walls of certain bacteria, and a substance related to the sugar L-glucose exists in streptomycin.

Explorers on an earthlike planet prepare to determine the local biochemical patterns, to learn whether life from Earth will pose a threat. (Courtesy Don Dixon)

have to do would be to compare biochemical standards. Since such standards are apparently a matter of chance, there would be three chances in four of his *not* having the same standards, both in proteins and in sugars, as the life of the new planet. His people then could land and build at least a small settlement or outpost in calm assurance that even if some of their animals were to get loose, there could be no chance of devastation of indigenous species. A stock animal that wandered into the forest would not run wild and tear up young seedlings. Instead, it would be found dead of indigestion.

We have sought to discover reasons why starfarers might not visit Earth, but in all honesty we have time and again come up with good reasons why they might. Then we recall the lack of evidence for civilizations not of our making, for fossil species from evolutionary patterns foreign to our own. And we must wonder.

Therefore, our exploration of the Fermi Paradox must end in mystery. The simple answer is to agree with Michael Hart: We have tried to explain the absence of extraterrestrials; we have reached no convincing answer. We then can only conclude that they do not exist, that we are indeed alone, unique.

The mind recoils from this. To be alone, with none like unto ourselves, is a status few would willingly accept. Yet we may find that this is the cup that has been vouchsafed to us, and from which we cannot fail to drink.

Today these questions are esoteric, academic. Tomorrow, as their full importance sinks in, they may prove to be the foundation for new science, new philosophy. The question of our relation to the Galaxy, the possibility of our uniqueness may call forth the attention that past civilizations gave to the question of our relation with God, the possibility of salvation. As Europe's cities vied to erect cathedrals, so may future cultures build space colonies—and starships.

229

CHAPTER **13**

Perspectives on the Future

The subject of the future is a popular one, and there is no shortage of articles and books in this area. However, most such opinions have a curiously static quality. They assume that only a few aspects will change, such as population or the availability of energy; or they grind political axes by warning of doom in tones that hint of threats to human survival; or else they promote technical miracles to flow from some new invention. There rarely is a serious attempt to reflect the richness and diversity of any human experience, including that of going forward into the future. Most of all, few authors have come to grips with the contradictory nature of different trends. Yet inevitably a topic so complex as the future must be subject to many trends that will shift in importance and influence, alternately giving and withholding hope and bestowing that gift first upon one group of peoples, then another.

With these caveats, it is appropriate to offer what after all are my personal opinions. I think that humanity will advance and advance to a time of space colonies. I think there will be interstellar flight and that our descendants will make their mark upon the Galaxy. But while we do these things, we will face new and difficult times in which we will struggle with the effects of our past growth and progress.

What may be the most significant of these effects will involve nothing so straightforward as pollution or environmental despoilation, popular though these topics are. Indeed, these effects are properly described not as environmental at all, but as climatic.

As early as the 1930s there were scientists who appreciated that the growth of an industrial society might lead to changes in the weather. It was G.S. Callendar who pointed out that the burning of coal and other fossil fuels would lead to a buildup of carbon dioxide in the atmosphere. This CO_2 then would act to trap heat by the greenhouse effect, raising global temperatures. The years 1880 to

1940 were years of growth in industrialization, and meterological records showed the average temperature rose by nearly 1° F. in those years. It was quite tempting to suggest that the two trends were related.

A degree of temperature matters little in a day's weather, but as a change in a global average, it means a lot. Reid Bryson, director of the University of Wisconsin's Institute for Environmental Research, has shown that a drop in the global average of 1.8° shortens the growing season for crops by two weeks; in addition, the cooler growing days allow less growth. The result is a falloff in crop yields of 27 percent. A drop in the global average of 4.3° would cut crop yields by 54 percent and would effectively wipe out many agricultural regions.

After 1940 world temperatures turned downward, while industry (and fossil-fuel use) grew apace. Apparently things were not so simple; some scientists suggested that there was actually a net removal of CO_2 due to expansion of agriculture. Evidently the scientists needed more data. In 1958 Charles D. Keeling, of Scripps Institute of Oceanography, began regular monitorings of CO_2 in the atmosphere atop the Hawaiian mountain Mauna Loa, which was far from any industrial center. Keeling's associates also undertook similar measurements at the South Pole. As the 1960s progressed, other scientists succeeded in devising means to determine past global temperatures, to understand how CO_2 would be interchanged with the oceans, and to calculate with better accuracy how an increase in CO_2 would lead to a rise in world temperatures.

By the 1970s the time was ripe for important advances. Based on studies of oxygen isotopes in ice cores taken from Camp Century, Greenland, W. Dansgaard and other glaciologists found a record of fluctuations in mean temperatures over past centuries. These fluctuations were quite adequate to account for the observed changes; this meant that human activities had not yet become a dominant influence. The news from Hawaii was less sanguine. Keeling showed that from 1958 to 1976, atmospheric CO_2 increased by 5.4 percent, from 314 to 331 parts per million. That would be enough to raise world temperatures above their pre-industrial values by at least half a degree, according to such atmospheric scientists as NASA's Ichtiaque Rasool. Moreover, this CO_2 buildup had followed closely the rise in burning of fossil fuels and could be understood if half the man-made CO_2 stayed in the atmosphere while the rest dissolved in the oceans. Indeed, a number of oceanographers showed that this occurrence was quite reasonable, though the consensus was that the world's forests also helped by absorbing some of the CO_2 in their growth.

With that, anyone reasonably acquainted with world industrial trends could show that by the early decades of the next century, CO_2 levels were likely to rise above 500 parts per million, close to double the pre-industrial value. According to the most widely accepted climatic models, the attendant rise in world temperatures would be about 4°, which would make Earth's climate warmer than at any time in the last thousand years.

If the resultant temperature rise was consistent throughout the planet, we might well say it would be pure delight. The last time global temperatures were even a couple degrees higher than today was around the year 1000, when Viking settlers were able to do farming in what was then the aptly named Greenland. The cited atmospheric models predict increased rainfall with the added CO_2, which with the warmer temperatures would mean longer growing seasons, more vigorous growth of crops. The expectation that temperatures would stay at 1940–1950 levels led Soviet planners to seek to make cropland of the ''virgin lands,'' marginal croplands in central Asia, in a project which failed dismally. A few extra degrees could make a lot of difference to the agriculture of the Soviets and of Canada, too.

Indeed, to the extent that growing CO_2 levels may enhance crop yields, the world's governments would welcome this effect and resist efforts to cut down on their CO_2 releases.

That could prove their undoing. If temperate-zone climates were to warm by some 4°, then these same atmospheric models predict that polar temperatures would rise by up to 16°. That could well prove sufficient to change the temperature structure of the oceans, thus causing a major rise in sea level.

The temperate-zone oceans consist of a surface layer of relatively warm water, a zone of intermediate temperatures that decrease with depth, and a deep cold layer. It is this way because at cold polar regions, icy water descends to the ocean bottom and migrates at depth toward the equator. Any warming of the polar regions thus could warm the oceanic abysses as well, which now are only a few degrees above freezing; eventually the whole of the oceans would grow warmer. The oceans contain some sixty times as much CO_2 as the atmosphere, and each degree of warming could release enough CO_2 to increase the atmospheric content by 3 percent, leading to still further warming.

The syndrome of human-generated CO_2, polar warming, deep ocean warming, and release of oceanic CO_2 could bring rapid, unplanned changes, which would cause severe global difficulties. Moreover, there are enough fossil fuels available to produce a rise in CO_2 levels not merely twofold, but tenfold. This syndrome could well result in that often-predicted and much-feared event, the melting of the polar caps. Depending on how extensive the melting and how strongly it would affect the mile-deep glaciers that cover Greenland and Antarctica, the rise in world sea levels could exceed a hundred feet.

It all could happen quite rapidly, too. It has happened before. The last major advance of the Ice Age glaciers was some seventy thousand years ago. For nearly sixty thousand years the ice sheets endured, a mile deep where Chicago stands today, extending south to the Carolinas, burying much of Europe beneath their massive frigidity. Then about fifteen thousand years ago, the glaciers, those sturdy and interminable institutions, began to melt; and when they did, how quickly they went! As the geologist Cesare Emiliani has noted in a 1975 *Science* article:

> The concomitant, accelerated rise in sea level, of the order of [a foot] per year, must have caused widespread flooding of low-lying areas, many of which were inhabited by man. We submit that this event, despite its great antiquity in cultural terms, could be an explanation for the deluge stories common to many Eurasian, Australasian, and American traditions. . . . This age [of 11,600 years] coincides, within the limits of all errors, with the age assigned by Plato to the flood he describes.

That earlier glacial melting lifted world sea levels by close to three hundred feet. If the sea level were to rise by a foot each year, many seaports and coastal areas could stay dry only by building seawalls and dikes on a massive scale. Many parts of the world might seek to become latter-day versions of Holland, and it is an interesting question how successful they would be. The drama of Noah's Ark could be played out in the flooded streets and avenues of Manhattan.

The situation is even more complex than this and thus potentially even more severe. Until recent years it was believed that a worldwide growth of living plants was serving to take up some of the excess CO_2. This belief made it easier for oceanographers to account for the flow of CO_2 into the oceans, since it limited the flow to levels consistent with their understanding of the oceans. However, in recent years such biologists as George M. Woodwell, director of the Ecosystems Center at Woods Hole, Massachusetts, have emphasized that the worldwide clearing of forests represents an additional source of CO_2. This new source may be as large as that from fossil fuels. Trees chopped down are usually burned or left to decay; in either case there is CO_2 release.

This discrepancy has caused a conflict between biologists and oceanographers. The biologists have pointed to much excess CO_2 from man's clearing of forests and have noted that the total CO_2 production, from forests and fossil fuels combined, far exceeds the increase in atmospheric CO_2 as measured by Keeling. They have suggested that the excess winds up in the oceans. The

Low-lying lands such as Florida would be flooded by a rise in sea levels. (Courtesy NASA and Lunar and Planetary Institute)

oceanographers have offered much experimental evidence to show that they understand CO_2 transport into the oceans and have insisted that in no way could the oceans accept so much excess as the biologists would wish. And while the scientific debates continue, the CO_2 buildup in the atmosphere proceeds apace.

The consequences of this debate will affect us all. If the biologists are right, then the sea can absorb more CO_2 than is now believed, and it may be a long, long time before there is a climatic change. If the oceanographers are right, the buildup of atmospheric CO_2 will continue with increasing speed, and the current rapid clearing of tropical rain forests may make the buildup proceed even faster. Until there is resolution of this issue, there will be no chance that scientists will be able to issue a clear warning of danger, much less that any warning would be heeded.

Without knowing it, humanity has entered into a vast global experiment with consequences that are poorly understood. These consequences could include shifts in patterns of wind and rainfall or of ocean currents such as the Gulf Stream. It is quite possible that humanity will inadvertently cross a critical threshold and trigger a rapid climatic warming. If unpleasant effects arise, they will not be quickly reversed and may not be reversed at all. Animals today generally are adapted to relatively cool conditions, and climatic change could mean vast shifts in Earth's patterns of life. A rise in sea levels could go beyond the capacity of societies to adjust easily.

These changes may take place during the very decades when space colonization could become significant, and it is tempting to imagine how activities in space may alleviate or even remove the climatic effects. To keep the CO_2 buildup under control, the world would have to limit its burning of fossil fuels and restrict the clearing of forests. Yet these policies would mean social and economic upheavals virtually as severe as those from the CO_2. The best solution would be for the world's peoples voluntarily to adopt new energy sources as replacements for fossil fuels, and this may yet come about.

One of these energy sources could well be the power satellite. It is one way to get energy without CO_2, and while it is not the only one, it certainly will see intensive interest. If it proves the least costly and most readily expanded energy source, as it may, then the powersat may sweep all before it.

Space colonies then will have an inevitable distinction indeed, for they will become centers for the building of the next century's valued energy plants. Far from being a matter for speculation, as it is today, space colonization will attract vast resources and efforts. It would be the history of Alaska all over again. At the time of Alaska's purchase (for two cents an acre) in 1867, Secretary of State William Seward was accused of wastefulness for buying ''Walrussia,'' or ''Seward's Folly.'' Time passed, and the day came when the nation came to look to that very land as an important source of oil.

In the decades to come, shortages of oil may delay but will hardly prevent a growing worldwide surge toward industrialization. The worldwide demand for higher living standards is so pervasive, the possible means of obtaining needed energy and resources so numerous that such rising living standards surely will be achieved. A growing accumulation of atmospheric CO_2 can only result from increased use of coal, oil shale, and (to the extent available) petroleum, and it is simply wrong to regard this as a bad thing, a form of pollution. What it will also mean will be rising worldwide affluence.

To speak of worldwide affluence may seem a bad joke, for today we are 4.2 billion people, most of them poor. Our tribe increases by some 70 million each year. There is no denying that today poverty is widespread, thoroughgoing, pervasive, and increasing year by year in the number of people it touches. The one thing it is not, or need not be, is interminable.

A rise in the sea levels could mean the abandonment of such cities as Los Angeles. This scene is set in the year 2200: What once was that city now is marshland, and the Santa Monica Freeway has fallen into ruin. (Courtesy Don Dixon)

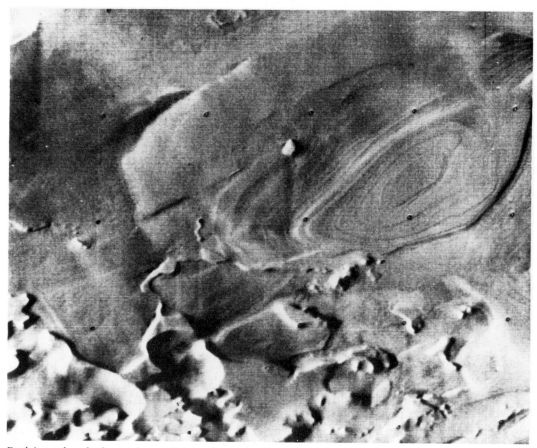

Earth is not the only planet to have undergone climatic change. These concentric zones of light and dark material were laid down on Mars during alternating times of snowy and dry climate. (Courtesy Jet Propulsion Laboratory)

Worldwide affluence means the growth of a worldwide middle class. The definition of middle class varies from culture to culture even within a given country; thus autos, a virtual necessity even for many poor people in California, can be happily dispensed with at all income levels in Manhattan. Nevertheless, a broad measure of middle-class status would include ownership of autos and telephones and use of electricity and crude oil. Population specialist Nathan Keyfitz, of Harvard, has found that with such a measure the world middle class numbered some 500 million in 1970 and increased to perhaps 600 million by 1975. Significantly, the same index gives a total of only some 200 million in the world middle class in 1950. During those years, 1950 to 1975, world population grew from 2.5 to 4.0 billion, for an average annual increase of 1.9 percent. The average annual growth in the world middle class was 4.5 percent.

If these trends were to persist, then as early as the year 2025 a majority of the world's population, some eleven billion, would enjoy middle-class standards of affluence. It is against this backdrop that one may assess the prospects for China's ambitious program of ''Four Modernizations'' (of agriculture, defense, industry, science and technology), which aim to bring the standard of living of her billion people to Western levels by the year 2000. The aim is a growth rate of 7 percent per year and the fact that China is self-sufficient in oil will help. By contrast, Japan enjoyed a growth rate of some 15 percent from 1950 to 1973, and since that oil-embargo year still has averaged 8 percent or 9 percent.

The years 1970 to 1973 were unprecedented in economics. Agricultural production rose; the Green Revolution offered increased yields for many crops. Petroleum was so abundant and cheap that even Japan and the nations of central Europe, oil-poor and long accustomed to coal, switched to the new fuel. Resources were ample to support growth both of economies and of populations. Today, of course, the situation is rapidly changing. Rising prices for energy, for agricultural products, and for many industrial raw materials signal increasing pressure by today's people on the available resources. Thus, will the trends of that quarter-century continue; will such advances in worldwide affluence indeed materialize? In *Colonies in Space* I used the work of Alvin Weinberg to point out that with the important exceptions of hydrocarbons and phosphorus for agriculture, the most extensively used materials exist in extractable quantities which at 1968 rates of use would last for millions of years. This is not to say that today we do rely on the resources that contain these materials in greatest total abundance. On the contrary, we usually rely on resources that furnish needed materials at the lowest cost, even if they can do so for only a few decades. Given time, the economic system is resilient and can substitute common materials for scarce ones, but this replacement does not happen overnight. What is needed is not only invention and innovation, but the huge capital investments that serve to develop new industries to replace old ones.

The long-delayed rise of a U.S. shale oil industry is a case in point. The world's greatest reserves of hydrocarbons are not in Saudi Arabia. They are found in the Green River, Uinta Mountains, and Piceance Creek oil-shale formations of Wyoming, Utah, and Colorado; but as with solar energy, the needed technology and capital have only lately seemed available. There is irony when motorists in these states curse the gasoline situation as they form long lines at the pumps; it is uncomfortably similar to Arthur Clarke's fable of cavemen freezing to death atop a mountain of coal.

As many writers have stated, the most important resource is human ingenuity. It is this resource, far more than any other, which will determine the pace of growth of the world middle class by overcoming limits of existing material supply and opening the prospects for substitutions or replacements. The economist Simon Kuznets of Harvard, a Nobel laureate, has summed up the world's economics: "Lack of resources is not the cause of underdevelopment. It is underdevelopment that is the cause of lack of resources."

With this perspective, power satellites may or may not lead a world advance to prosperity, but they can be very much a part of the work of human inventiveness. It is this inventiveness that may bring new agricultural techniques, better climate forecasts, fusion, new contraceptives, improved electronics, more widespread use of motors in place of human labor or draft animals, more responsive governmental policies, icebergs as a source of fresh water, increasingly influential regional economic groupings of nations, space colonies, and much more.

If space colonies open up as a virtually limitless human milieu, it is quite possible that they will see vast immigrations. People tend to move to wherever their prospects and hopes will be brightest, and the day may come when tens and even hundreds of millions of people live in space. Still, that will probably take several centuries, and it is not clear that rocket transport will ever be relatively less costly than the sea travel that brought America's settlers. In the course of a century of immigration,

237

some thirty million people came to our shores—a total that today is exceeded by five *months* of growth in today's world population. It will be at the very least a long, long time before a space nation will offer the prospect of siphoning off Earth's growth in numbers.

No, the problem of population, which limits economic growth rates and delays improvements in living standards, probably will be faced and met on this planet. In anticipating a long-term future for humanity, whether on Earth or in space, one of the most fascinating possibilities is that of population shrinkage, not growth; and today's developed nations may show the way.

To achieve zero population growth, the women of a nation must have an average of 2.1 babies each; this is the so-called "replacement fertility." Of these, more will be boys than girls, while some will not live to reproductive age; that is why the figure is 2.1, not 2.0. Since census bureaus deal with births year by year, rather than over the decades of a woman's childbearing years, there is the yearly "fertility rate": the number of children 1,000 women would bear if they had children through all their fertile years at that yearly rate. A fertility rate of 2,100 then is the same as replacement fertility, if kept up for enough years.

In America, the post-World War II years saw the Baby Boom, but the past decade has seen quite a different trend. The fertility rate peaked at 3,724 in 1957, but subsequent years saw it fall sharply. In 1970 it was 2,447; in 1972, 2,025, which for the first time was below replacement level. Since then it has fallen further. It was 1,900 in 1973, about 1,800 in 1978, and still is trending down. According to Charles Westoff, director of Princeton University's Office of Population Research, the rate is likely to go to 1,700.

Since there are a lot of young women in America, our population will keep rising for a while, albeit slowly, despite these low fertility rates. In Europe the situation is quite different. There the fertility rate in a number of countries may reach 1,500 by 1986 and stay there. In 1978 Westoff pointed out that both Germanys, the United Kingdom, Austria, and Luxembourg already were experiencing population declines. In 1979 Lester Brown, president of the Worldwatch Institute, added Belgium and Sweden to the list. He also suggested that by 1985 this roster might grow to include France, Italy, Japan, the Soviet Union—and the United States.

Why should this be? Better contraceptives have helped, but the U.S. fertility rate was between 2,100 and 2,200 during the Depression, long before the Pill. Much more important than contraceptives is the will to use them. Throughout the developed world there have been trends for women to receive more education, to marry later and to have the freedom to divorce, to spend at least part of their careers working outside the home, and to control the number and spacing of their babies. All these are powerful influences, which run counter to the traditional role of a wife who marries early and has lots of kids. Instead of motherhood being a woman's prime goal and role, it is increasingly only one of a number of interesting things she may do with her life.

The Baby Boom was actually an upward blip, albeit a significant one, on a 150-year trend to lower U.S. birth rates. In 1820 the birth rate was 55 births per 1,000 of population; by 1900 it was 32, and during the 1920s skidded to 21. During the Baby Boom it jumped up again, peaking at 27, but then fell back. By 1975 it was at an all-time low of 14.8. Still, this rate was not the lowest. In France in 1976 the birth rate was 13.6; in East Germany in 1975, 10.6. Even China apparently had the very low birth rate of 14.0 in 1975.

So it appears that these trends will continue and will touch more and more nations. In undeveloped economies children may be valued as economic assets; they can help with tilling the fields and can provide for the parents in old age. With the growth of cities, the advent of more sophisticated agriculture, the rise in educational levels, children cease to be assets and become economic liabilities. This, plus advances in the status of women, serve as powerful incentives in reducing population growth.

In today's advanced nations, some 30 percent of young women will likely go through their childbearing years without having any children. This means that once the population levels off, it will turn downward with increasing speed; for if 30 percent are childless, to reach the replacement level of 2,100 births per 1,000 women the remaining 70 percent would need three children each. As Charles Westoff has put it *(Marriage and Fertility . . .),*

> There is no magical quality in the 2.1 fertility rate that just maintains replacement. No one has yet discovered any [self-acting] mechanism that will automatically regulate a society's reproduction to keep it at the replacement level. If fertility continues downward, the ultimate prognosis is negative population growth: declining population size and much older populations. . . . The aging of the population will make it increasingly difficult to arrest the demographic momentum of the decline.

As these new population trends gain strength, it will become evident that many nations have embarked on another vast experiment. This one will test whether it is possible to have both replacement-level fertility and high levels of status and independence for women. It may be instead that even with elaborate pro-natalist programs, it still will prove quite difficult to prevent population declines. After all, baby-making is not only a highly personal choice; it also is par excellence the sort of thing of which a couple may say, "Let John and Marsha do it."

What's more, small families can live better and enjoy greater affluence. The difference between two children and three amounts to a 25 percent gain in income per person, which is more of a rise in living standards than a working man may achieve with ten years of raises. The future then could well be one of limited populations and, in many parts of the world, diminishing numbers of people, but all living better and better.

There is no obvious limit to this. If the world population peaks out at ten billion and thereafter declines at an average of one-half percent per year, the passage of fourteen hundred years will see a reduction to ten million. It may be hard to accept, but the mathematics are plain. And as long as declining populations pose no evident discomfort to people's lives, there will be little pressure to increase the birth rate. By the year 3000, the U.S. population could be down to the levels of a few hundred thousand that prevailed before 1492, but instead of living like Indians, each of these favored few would live at least as well as Jacqueline Onassis.

This may be the world of the space colonies. This world, within which the colonies will flourish and expand, may be one of man-made climatic changes, of shrinking populations (or, in Africa, Asia, and Latin America, populations which are in the process of peaking out), and of a growing affluence produced by the work of human ingenuity. In such a world, powersats and space colonies need not be rejected or dismissed merely because they appear to be far-out ideas; the world will need many far-out ideas to maintain its recent trends toward prosperity. Nor may such space projects be left to languish for want of needed capital or support, for vast flows of capital will come forth for the most promising prospects and the best-founded new programs. The building of powersats and the colonization of space will be weighed on their merits.

A trend to limited or shrinking populations may well mean that the human race, despite a significant reach into space, would still keep Earth as its center. The space colonies might be the new

America, but it is worth remembering that this continent has never held more than 7 percent of the world's people, despite centuries of growth and immigration. In his famous 1974 article in *Physics Today,* ''The Colonization of Space,'' Gerard K. O'Neill of Princeton University pointed out that the resources of the asteroids alone could support twenty thousand times the present world population at very high standards of living. Yet if the colonies follow the precedent of America, they may never support more than a few percent of humanity. In a world that increasingly may emphasize quality of life over quantity in numbers, the colonies could lead the way.

One area in which the colonies can take this lead will be in the introduction and increasing use of robots. The initial use of robots in construction activities will mean that they will be an integral part of life, and colonists will be inclined to find new opportunities to use them. If the population of humans is to shrink while that of robots is to grow, might the day come when they would become the dominant species?

In Chapter 2 I mentioned and then put aside the question: Could silicon-based life grow out of carbon-based life? As the reader may have guessed, what I meant was the origin of an intelligent species manifesting life as we know it, which goes on to invent computers and robots and then loses control of its own inventions. If robots are to proliferate, might the orbiting colonies prove a variation on the theme of monsters from space that take over from humans?

Certainly, computers and robots can be made to superficially mimic life, even intelligent life. However, they lack one essential element, which is life itself.

The basic unit of life is the cell, and one-celled plants and animals are quite common. Under the microscope they swim and move to and fro. They must ceaselessly carry out a metabolism, taking in some substances and excreting others; they continually require energy. They recognize food; they avoid less hospitable surroundings and seek the more hospitable. They can exchange genetic material with others of their kind, through the process known as conjugation, and at length they will fission and reproduce. Theirs is a continual struggle to maintain their molecular complexity, to avoid being broken up into simpler and more stable compounds. Theirs is the law: Grow, or die.

The basic unit of a computer is a transistor or similar circuit element, a simple structure of silicon. It exists in one of two electronic states, depending on the flows of current it has received—and that is all. It conducts no metabolism. It seeks no food, nor does it seek electrons. It can exist amid rapid switchings between states or in quiescence—it is ambivalent. It need do nothing to prevent death or decay, for it is already in a simple chemical state, which it can sustain throughout the ages.

In ages past men erected statues of stone that were cunningly wrought, and, carried away by the skill of their work, declared that the statues were gods, with the power and influence of a living king. The art of building computers and robots is skillful indeed, yet it does not alter the nature of their materials or change them from nonlife to life. A device like the Speak and Spell is indeed a wonder; it is akin to teaching a stone to speak. Yet the stone remains a stone. It is not a man, nor is it a new form of life.

Yet if it is not through robots that space colonies may devise a next step beyond man, still in these colonies the successor to *H. sapiens* indeed could be created. The colonies will be quite diverse, and it will be a common thing for specialized groups of people, with distinct customs and interests, to go off to the far reaches of the asteroid belt and establish space colonies of their own. Perhaps among these groups there will be some dedicated to developing an advanced or improved type of human being.

To say that this idea is unpopular, that it conjures up visions of every horror from Frankenstein's monster to Nazi experiments, is to explain why the advocates of this goal might seek to live and work far from the broader community of Earth-people. The human culture will tolerate and even encourage much. Few customs or codes of law will appear unacceptable, and differences in people's appearances will carry less and less weight. What we have no experience with in modern times, what we have not faced since the Cro-Magnon gained dominance over the earlier Neanderthalers, is for an established human stock to face a genetically superior species.

By "genetically superior" I have in mind the Darwinian concept of competition between species, rather than any cultural patterns. As to what endowments, what innate ways of thinking, what physical changes could constitute transhumanity—that is a hard matter to address. It is difficult to say what the limits are to cultural changes, or to what degree such changes could fail to leave humanity equipped to meet future challenges. Yet we know that people have all too frequently suffered from mass delusion and self-delusion, from tendencies to be carried away by their emotions at the expense of their reason, and from difficulty in adopting critical or skeptical modes of thought or in weighing present desires against the more significant needs of the future. Are these merely cultural shortcomings that could be overcome with better forms of education? Or will some group of experimenters link these features to structures of the human brain, which can then be rewired?

The question of human conformity, the urge to follow and belong to a group, may be important here. Harvard's Edward O. Wilson, founder of the science of sociobiology, has written of "conformist genes," suggesting that conformist tendencies are inherited, not learned. Paul MacLean of the National Institutes of Mental Health, whose work formed much of the basis for Carl Sagan's *The Dragons of Eden,* has traced the roots of conformist behavior to the deep-seated structures of the brain known as the R-complex. He has pointed out that Einstein was a nonconformist and a loner and has suggested that "some individuals may become creative because of a constitutional incapacity for imitation . . . a defect of the nervous system that might interfere with the intercommunicative [group-feeling] process." Such a person could talk with members of a group, and join in its activities, yet would not necessarily feel himself to be a part of that group. A genetically superior humanity then might simply be one in which the R-complex has been modified or weakened. This could come through editing the genetic code, which contains the specifications for a human being.

I do not know whether anyone will ever determine the complete sequence of the nucleotide bases that serve to specify a human being, but it may happen. For several years the means have existed to do precisely that, at least in principle. All living beings have their nature coded in long molecules of the genetic material, DNA. The code consists of sequences of the molecules adenine, thymine, cytosine and guanine, denoted A, T, C, G; any triplet of these so-called nucleotide bases then codes for or corresponds to a specific amino acid. A complete nucleotide sequence is a long string of these bases. It is like an encyclopedia of words all run together, written with an alphabet of four letters.

Thus far, genetic codes have been worked out only for such simple creatures as viruses. For instance, Frederick Sanger and his associates at Cambridge University have found that a bacterial virus designated ϕx174 has a nucleotide sequence of 5,375 bases that are grouped into nine genes, which in turn code for the amino acid sequences of nine different proteins. The complete nucleotide sequence for ϕx174 fills a page of ordinary type. A typical bacterium would need two thousand such pages to represent its code; a man would need a million pages.

The working-out of such sequences has involved techniques closely allied to those that actually serve in the manipulation of genes and the revising or editing of genetic codes. It is not appropriate here to go into the controversy that these techniques have spawned in recent years. However, these methods permit the extraction of sequences of DNA from one species and their transplantation or splicing into the DNA sequence of an entirely different species. These methods, known collectively as "recombinant DNA," have allowed biologists to create entirely new forms of bacterial life. Such bacteria have already been successfully designed to manufacture the brain hormone somatostatin, and there recently have been major advances in transplanting rat genes into bacteria so as to cause them to produce insulin.

It is too early to say what will be the ultimate impact of this control over the ultimate sources of life. For now, we must imagine that these techniques, or their even more sophisticated descendants, have allowed a group of researchers in a space colony to evolve a genetic sequence which, when incorporated in a fertilized human ovum, leads to the first of the species *Homo transsapiens.* In that distant colony, guided by the researchers, the transhumans could then assemble their gene combinations and increase in number.

And when they venture forth to face a solar system dominated by *H. sapiens,* what response will they find? Will they be greeted in friendship, welcomed to an honored place in the family of man? Or will they meet fear and hostility, and the certainty that however few their number, they will be regarded with the awe and fear with which the Neanderthalers must have faced the Cro-Magnon?

If they are to be treated as outsiders, relegated to remote regions of the Solar System rather than be granted the opportunity to make good their competitive advantage, then these transhumans might emigrate rather than accept such restrictions. They could not emigrate to any part of Earth, of course, nor to any other region of nearby space. Instead, their port of call would be—the stars.

It is no new thing to write of star flight, star colonization; what has not been so evident has been a motive. It cannot be commerce, nor population pressure, or even wanderlust or adventure. To speak of starfarers as adventurers is in a sense a contradiction in terms. They would have to spend decades or centuries under disciplined restrictions while en route before their descendants could establish their settlements. And in all but a vanishingly small number of cases, the settlements would be space colonies built from asteroidal materials, differing only insubstantially from what they might have built in unsettled regions of our own Solar System.

Yet what if the motivation is not adventure, but opportunity? What if the emigration is a necessity, if the starfarers are to avoid hostility or severe restrictions on their actions? On Earth oppressed peoples have commonly sought freedom in new lands beyond the seas. This oppression has stemmed from cultural differences; how much more significant may the differences be if they are not cultural, but genetic.

Still, the question of a motive is one that may resolve itself in time. The idea of transhumanity achieved through recombinant DNA is no more than a speculation; far more so is the notion of transhumans becoming starfarers in order to escape persecution. After all, recombinant DNA today is no more than the latest technique in the human quest for knowledge. In the early 1800s the latest technique was electricity, which Galvani had used to cause frogs' legs to twitch. This was the discovery that inspired Mary Wollestonecraft Shelley to conceive the legend of a new form of human invented by her fictional medical student, Dr. Frankenstein. To write of transhumanity fleeing to stellar space may be simply a fable, a suggestion of the character of a world that could send ships to the stars.

In the end, we can only say this: If space colonization goes forward, in time people will have both the ability and the means to seek the stars. We have been in such situations before, and we have seldom disdained to take advantage of the new opportunities. It has lately been fashionable to say that all we need is on Earth, that there is no reason save mere vainglory to venture into space. The whole of

Will transhumanity arise in a space colony? (Courtesy Don Davis)

this book argues against this. In the future it may be said that all we need will be in the Solar System, that there will be no reason to seek the stars. Perhaps that view will in time also pass away, even if for reasons that today are obscure.

We do know that if humanity persists and endures, in time we will come face to face with the evolution of our sun. In a few billion years its slow brightening will speed up as it swells into a red giant. Earth then will be uninhabitable, as will the inner regions of the Solar System. Yet there will be other and more clement stars to which our descendants may wish to migrate. Certainly, a society that has developed space flight and space colonization will have the advantage of never thereafter having to stand hostage to fortune.

There still is the question posed in the last two chapters: Are we alone? The answer is there to be found. It exists somewhere out among the stars, and it will be our spacefaring descendants who will learn it.

Descendants of space colonists, heirs to a world deeply influenced by these colonizations, will be touched and shaped by their milieu. It is that milieu, what the poet Diane Ackerman has called the cosmic overwhelm, which even today looms as one that will increasingly occupy our attentions in the decades, the centuries ahead. It will give terror and delight, loneliness and fond companionship, reverence and awe. It will be a challenge to face, a source of hope and wonder, a highway, a sea—yes, and a place of death.

There are people of islands and seacoasts, of rocky peninsulas where the surf murmers at night amid the salt spray. Their lives, their beliefs, their work all have entered our common heritage. The sea has governed and shaped their ways, fashioned their histories as surely as it has cut the ocean cliffs beside which they fish. Their thoughts and experiences have become part of our culture, and we are the richer for it.

So it will be with that vaster sea, that overwhelm, and with those who will build their homes afar, live their lives in the void. It will be thus with those adventurous ones, those creators of new things, those people of the future, who will venture toward distant suns.

BIBLIOGRAPHY

Among the books and conference proceedings that treat space colonies and power satellites in a general manner are the following:

Brand, S., ed. *Space Colonies*. New York: Penguin Books, 1977. Available from *CoEvolution Quarterly,* Box 428, Sausalite, California 94965.

Gray, J., ed. *Space Manufacturing Facilities (Space Colonies)*. Proceedings of the Princeton Conferences, vol. 1, 1974 and 1975; vol. 2, 1977; vol. 3, 1979. Available from American Institute of Aeronautics and Astronautics, 1290 Avenue of the Americas, New York, N. Y. 10017

Heppenheimer, T. A. *Colonies in Space*. Harrisburg, Pa.: Stackpole Books, 1977. Also available in paperback (Warner).

————, ed. *Power Satellites and the Industrial Use of Lunar Resources*. Progress in Aeronautics and Astronautics. New York: American Institute of Aeronautics and Astronautics, forthcoming.

Holbrow, C., and Johnson, R., eds. *Space Settlements—A Design Study,* 1977. Report of the 1975 NASA/ASEE summer study on space colonization. NASA SP-413. Available from Superintendent of Documents, U.S. Government Printing Office, Washington, D.C.

O'Neill, G. K. *The High Frontier*. New York: Morrow, 1977. Also available in paperback (Bantam).

————, ed. *Space-Based Manufacturing from Nonterrestrial Materials*. Report of the 1976 NASA summer study on space colonization. Progress in Aeronautics and Astronautics, vol. 57, 1977. Available from American Institute of Aeronautics and Astronautics, 1290 Avenue of the Americas, New York, N.Y. 10017.

————, ed. *Space Resources and Space Settlements*. Report of the 1977 NASA summer study on space colonization. NASA SP-428, 1979. Available from Superintendent of Documents, U.S. Government Printing Office, Washington, D.C.

Van Patten, R. A.; Siegler, P.; and Stearns, E. V. B. eds. *The Industrialization of Space*. Proceedings of the 23rd Annual Meeting, American Astronautical Society. *Advances in the Astronautical Sciences,* vol. 36, 1978. Available from Univelt, Inc, P.O. Box 28130, San Diego, Calif. 92128.

In addition, for current information in these areas, there are a number of magazines and newsletters: *L-5 News* (monthly magazine of the L-5 Society, 1620 N. Park Ave., Tucson, AZ 85719); *Subscribers' Newsletter* (Space Studies Institute, Box 82, Princeton, NJ 08540. The Institute's primary function is the support of research); *Spaceworld* (magazine published ten times per year by Palmer Publications, Inc., Amherst, WI 54406); *Space Age Review* (published monthly by Space Age Review, 355 West Olive Avenue, Sunnyvale, CA 94086); *INSight* (monthly magazine of National Space Institute, Suite 306, 1911 N. Fort Myer Drive, Arlington, VA 22209).

The following references have also been used:

CHAPTER 1

Abt, H. A., "The Companions of Sunlike Stars." *Scientific American,* April 1977, pp. 93–104.

Cameron, A. G. W. "The Primitive Solar Accretion Disk and the Formation of the Planets." *In The Origin of the Solar System,* edited by S. Dermott. New York: Wiley, 1978, pp. 49–74.

Goldreich, P., and Ward, W. R. "The Formation of Planetesimals." *Astrophysical Journal,* 183 (1973): pp. 1051–61.

Heppenheimer, T. A. "On the Formation of Planets in Binary Star Systems." *Astronomy and Astrophysics,* 65 (1978): pp. 421–26.

———. "Secular Resonances and the Origin of Eccentricities of Mars and the Asteroids." *Icarus,* in press.

Kirshner, R. P. "Supernovas in Other Galaxies." *Scientific American,* December 1976, pp. 88–101.

Schramm, D. N., and Clayton, R. N. "Did a Supernova Trigger the Formation of the Solar System?" *Scientific American,* October 1978, pp. 124–39.

Toomre, A. "Theories of Spiral Structure." *Annual Review of Astronomy and Astrophysics,* 15 (1977): pp. 437–78.

Woodward, P. R. "Theoretical Models of Star Formation." *Annual Review of Astronomy and Astrophysics,* 16 (1978): pp. 555–84.

CHAPTER 2

Berkner, L. V., and Marshall, L. C. "On the Origin and Rise of Oxygen Concentration in the Earth's Atmosphere." *Journal of the Atmospheric Sciences,* 22 (1965): pp. 225–61.

Dole, S. H. *Habitable Planets for Man*. New York: Blaisdell, 1964.

"Evolution." Special issue, *Scientific American,* September 1978.

Hart, M. H. "The Evolution of the Atmosphere of the Earth." *Icarus* 33 (1978): pp. 23–39.

———. "Habitable Zones About Main Sequence Stars." *Icarus* 37 (1979): pp. 351–357.

———. "Was the Pre-Biotic Atmosphere of the Earth Heavily Reducing?" *Origins of Life,* in press.

Molton, P. "Non-Aqueous Biosystems: The Case for Liquid Ammonia as a Solvent." *Journal of the British Interplanetary Society,* 27 (1974): pp. 243–62.

Newburn, R. L., and Gulkis, S. "A Brief Survey of the Outer Planets Jupiter, Saturn, Uranus, Neptune, Pluto, and Their Satellites." *Space Science Reviews,* 3 (1973): pp. 197–271.

Rasool, S. I., and de Bergh, C. "The Runaway Greenhouse and the Accumulation of CO_2 in the Venus Atmosphere." *Nature,* 226 (1970): pp. 1037–39.

Sagan, C. and Mullen, G. "Earth and Mars: Evolution of Atmospheres and Surface Temperatures." *Science,* 177 (1972); 52–56.

CHAPTER 3

Baker, W. A., and Tryckare, T. *The Engine Powered Vessel.* New York: Crescent Books, 1965.

Bekey, I., and Mayer, H. "1980–2000: Raising Our Sights for Advanced Space Systems," *Astronautics and Aeronautics,* July/August 1976, pp. 34–63.

Bier, M. "Space Bioprocessing—Status and Potentials." In *Future Space Programs,* hearings before the Committee on Science and Technology, U.S. House of Representatives, January 1978, pp. 300–15.

Covault, C. "Materials Processing Stress Questioned." *Aviation Week & Space Technology,* July 10, 1978, pp. 46–47.

———. "Platform Designed for Numerous Uses." *Aviation Week & Space Technology,* June 19, 1976, pp. 67–73.

Fordyce, S. W. "Communications Payloads for Geostationary Platforms." AIAA paper 78–1695, American Institute of Aeronautics and Astronautics, 1978.

Guillemin, R. "Peptides in the Brain: The New Endocrinology of the Neuron." *Science,* 202 (1978): pp. 390–402. See also *Science,* 200, articles beginning on pp. 279, 411, and 510.

Hotz, R., and Robinson, C. A. *Particle Beam Weapons.* Reprint of articles from *Aviation Week and Space Technology* May 2, 1977; Oct. 2, 9, 16, 1978; Nov. 6, 13, 1978. New York: McGraw-Hill, 1978.

Ley, W. *Rockets, Missiles, and Space Travel.* New York: Viking, 1957.

"Materials Processing in Space." Washington: National Research Council, 1978.

Mostert, N. *Supership.* New York: Alfred A. Knopf, 1974.

Pugsley, Sir A. *The Works of Isambard Kingdom Brunel: An Engineering Appreciation.* Reviewed in *Scientific American,* April 1977, pp. 144–46.

Safranov, V. S. *Evolution of the Protoplanetary Cloud and the Origin of the Earth and Planets.* Moscow: Nauka, 1969. NASA TT F–677, 1972.

Simpson, C. *The Lusitania.* New York: Little, Brown & Co., 1972.

"Space Industrialization—An Overview." Science Applications, Inc. Huntsville, Alabama 35805, April 15, 1978. Contract NAS8–32197; SAI Report 79–602–HU.

Wade, N. "Charged Debate Erupts Over Russian Beam Weapon," *Science,* 196, (1977): pp. 957–59. See also *Science,* 196, (1977): pp. 407–8.

Wolbers, H. L., and Shepphird, F. H. "Geosynchronous Information Services Platforms in the Year 2000." AIAA Paper 78–1636, American Institute of Aeronautics and Astronautics, 1978.

CHAPTER 4

Abelson, P. H. "Absence of U. S. Energy Leadership." *Science,* 189 (1975): p. 11.

———. "Energy Conservation Is Not Enough." *Science,* 196 (1977): p. 1159.

———. "How Much More Oil? *Science* 198 (1977): p. 451.

Covault, C. "Views Change on Power Satellite Work." *Aviation Week & Space Technology,* July 17, 1978, pp. 42–46.

Flower, A. R. "World Oil Production." *Scientific American,* March 1978, pp. 42–49.

Gordon, R. L. "The Hobbling of Coal: Policy and Regulatory Uncertainties." *Science* 200 (1978): pp. 153–58.

Hayes, E. T. "Energy Resources Available to the United States, 1985 to 2000." *Science* 203 (1979): pp. 233–39.

Holdren, J. P. "Fusion Energy in Context: Its Fitness for the Long Term." *Science* 200 (1978): pp. 168–80.

Hubbert, M. K. "Role of Geology in a Maturing Industrial Society." Resource Geology Seminar Series, California Institute of Technology, May 8, 1979.

Landsberg, H. H. "Coal: The New Swing Fuel?" *Science* 197 (1977): p. 9.

Marshall, E. "A Preliminary Report on Three Mile Island." *Science* 204 (1979): pp. 280–81.

Maugh, T. H. "Oil Shale: Prospects on the Upswing . . . Again." *Science* 198 (1977): pp. 1023–27.

Metz, W. D. "Mexico: The Premier Oil Discovery in the Western Hemisphere." *Science* 202 (1978): pp. 1261–65.

———. "Ocean Thermal Energy: The Biggest Gamble in Solar Power." *Science* 198 (1977): pp. 178–80.

Nagel, T. J. "Operating a Major Electric Utility Today." *Science* 201 (1978): pp. 985–93.

O'Leary, J. F. "Facing the Energy Facts." *Astronautics and Aeronautics,* February 1979, pp. 36–40.

Parkins, W. E. "Engineering Limitations of Fusion Power Plants." *Science* 199 (1978): pp. 1403–8.

Robertson, J. A. L. "The CANDU Reactor System: An Appropriate Technology." *Science* 199 (1978): pp. 657–64.

Steiner, D., and Clarke, J. F. "The Tokamak: Model T Fusion Reactor." *Science* 199 (1978): pp. 1395–1403.

Swabb, L. E. "Liquid Fuels from Coal: From R & D to an Industry," *Science* 199 (1978): pp. 619–22.

CHAPTER 5

Beichel, R. "The Dual-Expander Rocket Engine—Key to Economical Space Transport." *Astronautics and Aeronautics,* November 1977, pp. 44–51.

———. "Nozzle Concepts for Single-Stage Shuttles." *Astronautics and Aeronautics,* June 1975, pp. 16–27.

Hearth, D. P., and Preyss, A. E. "Hypersonic Technology—Approach to an Expanded Program." *Astronautics and Aeronautics,* December 1976, pp. 20–37.

Hearth D. P. *Outlook for Space.* NASA SP–386. Washington: U. S. Government Printing Office, 1976.

Jones, R. A., and Huber, P. W. "Toward Scramjet Aircraft." *Astronautics and Aeronautics,* February 1978, pp. 38–48.

Lay, B. *Earthbound Astronauts.* Englewood Cliffs, N.J.: Prentice-Hall, 1971.

"Liquid Rockets." In "Aerospace Highlights of 1978," *Astronautics and Aeronautics,* December 1978, pp. 60–61.

Rosen, M. W.; Pickering, W. H.; Lucas, W. R.; Sloop, J. L.; Schriever, B. A.; Phillips, S. C.; and Raborn, W. F. "Rocketry in the '50s." *Astronautics and Aeronautics,* October 1972, pp. 38–65.

Salkeld, R. "Mixed-Mode Propulsion for the Space Shuttle." *Astronautics and Aeronautics,* August 1971, pp. 52–58.

———. "Orbital Rocket Airplanes—A Fresh Perspective." *Astronautics and Aeronautics,* April 1976, pp. 50–52.

————. "Single-Stage Shuttles for Ground Launch and Air Launch." *Astronautics and Aeronautics,* March 1974, pp. 52–63.

Salkeld, R.; Patterson, D. W.; and Grey, J. eds. *Space Transportation Systems: 1980–2000.* AIAA Aerospace Assessment Series, vol. I, 1978, New York: American Institute of Aeronautics and Astronautics.

Scarboro, C. W. *Pictorial History of Cape Kennedy 1950–1965.* Indialantic, Florida: South Brevard Beaches Chamber of Commerce, 1965.

Simmons, H. "Space Shuttle: The Month that Was." *Astronautics and Aeronautics,* March 1979, p. 6.

Simpson, E. C., and Hill, R. J. "The Answer to the 'Engine Deficiency' Question." *Astronautics and Aeronautics,* January 1978, pp. 52–57.

CHAPTER 6

Britton, W. R. "Space Spider—A Concept for Fabrication of Large Space Structures." AIAA Paper 78–1655, American Institute of Aeronautics and Astronautics, 1978.

Covault, C. "Structure Assembly Demonstration Slated." *Aviation Week & Space Technology,* June 12, 1978, pp. 49–53.

DaRos, C. J.; Freitag, R. F.; and Kline R. L. "Toward Large Space Systems." *Astronautics and Aeronautics,* May 1977, pp. 22–31.

Disher, J. H. "Next Steps in Space Transportation and Operations." *Astronautics and Aeronautics,* January 1978, pp. 22–30.

Hagler, T. "Orbital Construction Demonstration Study Final Report." Grumman Aerospace Corp., Report NSS-OC-RP012, Contract NAS9–14916, June 1977.

Hagler, R., and Patterson, H. G. "Learning to Build Large Structures in Space." *Astronautics and Aeronautics,* December 1977, pp. 51–57.

"Large Space Structures—Challenge of the Eighties." Special section, *Astronautics and Aeronautics,* October 1978, pp. 22–59.

Muench, W. "Automatic Fabrication of Large Space Structures—The Next Step." AIAA Paper 78–1651, American Institute of Aeronautics and Astronautics, 1978.

CHAPTER 7

BORIS. Advertised in *Scientific American,* June 1978, p. 29.

Covault, C. "Tank Tests Validate Structure Assembly." *Aviation Week & Space Technology,* June 26, 1979, pp. 55–62.

"Electronics." Special issue, *Science,* March 18, 1977.

Holden, C. "The Empathic Computer." *Science* 198 (1977): p. 32.

"Microelectronics." Special issue, *Scientific American,* September 1977.

Miller, K. H., and Davis, E. "Solar Power Satellite Construction and Maintenance: The First Large Scale Use of Man-in-Space." AIAA paper 78–1637, American Institute of Aeronautics and Astronautics, 1978.

Olson, R. L.; Samonski, F. H.; and Miller, K. H. "Human Factors in Power Satellite Construction." In *Power Satellites and the Industrial Use of Lunar Resources,* edited by T. A. Heppenheimer, AIAA Progress in the Astronautical Sciences, American Institute of Aeronautics and Astronautics, forthcoming.

Shakespeare, W. *King Henry V,* act 4, scene 3, 1599.

"Space Station: A Guide for Experimenters." Space Division, Rockwell International Corp. SD 70–534, Contract NAS9–9953, October 1970.

Tewell, R. J. and Spencer, R. A. "Advanced Teleoperator Spacecraft." AIAA Paper 78–1665, American Institute of Aeronautics and Astronautics, 1978.

"Toward Space Robotics and Automation." Special section, *Astronautics and Aeronautics,* May 1979, pp. 16–46, 63.

CHAPTER 8

Austin, R. E. "Space Colonization by the Year 2000: An Assessment." NASA Marshall Space Flight Center, January 15, 1975.

Bock, E. "Lunar Resources Utilization for Space Construction." Final Report GDC–ASP79–001, General Dynamics Convair Division, Contract NAS9–15560, May 1979.

Greenberg, R.; Wacker, J. F.; Hartmann, W. K.; and Chapman, C. R. "Planetesimals to Planets: Numerical Simulation of Collisional Evolution."*Icarus* 35 (1978): pp. 1–26.

Hartmann, W. K. "Planet Formation: Mechanism of Early Growth."*Icarus* 33 (1978): pp. 50–61.

Heppenheimer, T. A. "Achromatic Trajectories and Lunar Material Transport for Space Colonization." *Journal of Spacecraft and Rockets* 15 (1978): pp. 176–83.

———. "A Mass-Catcher for Large-Scale Lunar Material Transport." *Journal of Spacecraft and Rockets* 15 (1978): pp. 242–49.

———. "Steps Toward Space Colonization: Colony Location and Transfer Trajectories." *Journal of Spacecraft and Rockets* 15 (1978): pp. 305–12.

Marshall, E. "Assessing the Damage at TMI"*Science* 204 (1979): pp. 594–96.

Rossin, A. D., and Rieck, T. A. "Economics of Nuclear Power." *Science* 201 (1978): pp. 582–89.

CHAPTER 9

Connor, T. "The Rose Garden Nobody Promised." *Saltwater Papaya* (Canal Zone College student magazine), Spring 1977, pp. 18–20.

McCullough, D. *The Path Between the Seas.* New York: Simon & Schuster, 1977.

Parsons, H. M. "What Happened at Hawthorne?"*Science* 183 (1974): pp. 922–32.

CHAPTER 10

Cole, D., and Cox, D. W.*Islands in Space.* New York: Chilton Press, 1964.

Cortright, E. M. *Exploring Space with a Camera.* NASA SP–168. Washington, D.C.: U. S. Government Printing Office, 1968.

Watson, K.; Murray, B. C.; and Brown, H. "The Behavior of Volatiles on the Lunar Surface." *Journal of Geophysical Research* 66 (1961): pp. 3033–53.

Wetherill, G. "Apollo Objects."*Scientific American,* March 1979, pp. 54–65.

CHAPTER 11

Calame, O., and Mulholland, J. D. "Lunar Crater Giordano Bruno: A.D. 1178 Impact Observations Consistent with Laser Ranging Results."*Science* 199 (1978): pp. 875–77.

De Lumley, H. "A Paleolithic Camp at Nice."*Scientific American,* May 1969, pp. 42–50.

Hartung, J. B. "Was the Formation of a 20-Km Diameter Impact Crater on the Moon Observed on June 18, 1178?" *Meteoritics* 11 (1976): pp. 187–94.

Horowitz, P. "A Search for Ultra-Narrowband Signals of Extraterrestrial Origin." *Science* 201 (1978): pp. 733–35.

Hynek, J. A. *The UFO Experience: A Scientific Inquiry.* Chicago: Henry Regnery, 1972.

Freeman J., and Lampson, M. "Interstellar Archaeology and the Prevalence of Intelligence." *Icarus* 25 (1975): pp. 368–69.

Jones, E. M. "Colonization of the Galaxy." *Icarus* 28 (1976): pp. 421–22.

———. "Further Calculations of Interstellar Colonization." Report LA–UR 79–738, Los Alamos Scientific Laboratory, New Mexico, March 1979.

———. "Interstellar Colonization." *Journal of the British Interplanetary Society* 31 (1978): pp. 103–7.

Martin, A. R., ed. "Project Daedalus." *Journal of the British Interplanetary Society,* Supplement, 1978.

Murray, B.; Gulkis, S.; and Edelson, R. E. "Extraterrestrial Intelligence: An Observational Approach." *Science* 199 (1978): pp. 485–92.

Sagan, C., ed. "Communication with Extraterrestrial Intelligence." Cambridge, Massachusetts: MIT Press, 1973, p. 186.

Sagan, C., and Page, T., eds. *UFO's—A Scientific Debate.* Ithaca, N.Y.: Cornell University Press, 1972.

"Space Watch's First Catch." *Time,* March 7, 1960, p. 80. See also *Time,* February 22, 1960, p. 14.

Sullivan, W. T.; Brown, S.; and Wetherill, C. "Eavesdropping: The Radio Signature of the Earth." *Science* 199 (1978): pp. 377–88.

Thomas, L. *The Lives of a Cell.* New York: Viking, 1974.

"UFO Encounter." *Astronautics and Aeronautics,* July 1971, pp. 66–70.

"UFO Encounter II." *Astronautics and Aeronautics,* September 1971, pp. 60–64.

Viewing, D. R. J.; Horswell, C. J.; and Palmer, E. W. "Detection of Starships." *Journal of the British Interplanetary Society* 30 (1977): pp. 99–104.

CHAPTER 12

Asimov, I. *The Left Hand of the Electron.* New York: Dell, 1972.

Ball, J. A. "The Zoo Hypothesis." *Icarus* 19 (1973): pp. 347–49.

Becker, C. L. *Modern History.* Chicago: Silver Burdett Co., 1958.

Carson, R. *The Sea Around Us.* New York: Signet Science Library, 1961.

Hart, M. H. "An Explanation for the Absence of Extraterrestrials on Earth." *Quarterly Journal of the Royal Astronomical Society* 16 (1975): pp. 128–35.

Jones, E. M. "The Zoo Hypothesis Revisited." In *Proceedings of the Southwest Regional Conference for Astronomy and Astrophysics,* edited by P. F. Gott and P. S. Riherd. Las Cruces, New Mexico, May 22, 1978.

Manchester, W. *The Arms of Krupp.* New York: Bantam, 1970, p. 50.

Papagiannis, M. D. "Could We Be the Only Advanced Technological Civilization in the Galaxy?" Astronomical Contributions of Boston University, Series II, no. 61, 1977.

Prelog, V. "Chirality in Chemistry." *Science* 193 (1976): pp. 17–24.

Viewing, D. R. J., and Horswell, C. J. "Is Catastrophe Possible?" *Journal of the British Interplanetary Society* 31 (1978): 209–16.

CHAPTER 13

Abelson, P. H. "Energy and Climate." *Science* 197 (1977): p. 941.

Ardrey, R. *The Hunting Hypothesis.* New York: Atheneum, 1976.

Broecker, W. S. "Climatic Change: Are We on the Brink of a Pronounced Global Warming?" *Science* 189 (1975): pp. 460–63.

Cohen, S. N. "The Manipulation of Genes." *Scientific American,* July 1975, pp. 24–33.

———. "Recombinant DNA: Fact and Fiction." *Science* 195 (1977): pp. 654–57.

Eaton, W. J. "Inflation May Curb Population Growth." *Los Angeles Times,* May 4, 1979, p. 4.

Emiliani, C. et. al. "Paleoclimatological Analysis of Late Quaternary Cores from the Northeast Gulf of Mexico." *Science* 189 (1975): pp. 1083–88.

Grobstein, C. "The Recombinant-DNA Debate." *Scientific American,* July 1977, pp. 23–33.

Holden, C. "Paul MacLean and the Triune Brain." *Science* 204 (1979): pp. 1066–68.

Kerr, R. A. "Carbon Dioxide and Climate: Carbon Budget Still Unbalanced." *Science* 197 (1977): pp. 1352–53.

Keyfitz, N. "World Resources and the World Middle Class." *Scientific American,* July 1976, pp. 28–35.

McLean, D. M. "A Terminal Mesozoic 'Greenhouse': Lessons from the Past." *Science* 201 (1978): pp. 401–06.

Siegenthaler, U., and Oeschger, H. "Predicting Future Atmospheric Carbon Dioxide Levels." *Science* 199 (1978): pp. 388–95.

Wattenberg, B. J. *The Real America.* New York: Capricorn Books, 1976.

Westoff, C. F. "Marriage and Fertility in the Developed Countries." *Scientific American,* December 1978, pp. 51–57.

Woodwell, G. M. "The Carbon Dioxide Question." *Scientific American,* January 1978, pp. 34–43.

Index

Page numbers in italics indicate illustrations.